PRESENTING

STATISTICS

A MANAGER'S GUIDE TO THE PERSUASIVE USE OF STATISTICS

PRESENTING
STATISTICS

A MANAGER'S GUIDE TO THE PERSUASIVE USE OF STATISTICS

Lawrence Witzling
Robert Greenstreet

University of Wisconsin — Milwaukee

WILEY

John Wiley & Sons

New York   Chichester   Brisbane
Toronto   Singapore

Drawings by J. Lackney

Library of Congress Cataloging in Publication Data:

Witzling, Lawrence, 1946-
Presenting Statistics: A Manager's Guide to the Persuasive Use of Statistics
/Lawrence Witzling, Robert Greenstreet.

p.   cm.
Bibliography: p.
Includes index.
ISBN 0-471-80307-3
1. Mathematical statistics   I. Greenstreet, Robert   II. Title.

| QA276.W583   1989 | 89-31521 |
| 001.4'226—dc20 | CIP |

Printed in the United States of America

10 9 8 7 6 5 4 3 2 1

Preface

In a wide range of professional pursuits, the presentation of statistical data to decision-making audiences is an integral part of successful practice. However, in many instances, the data may be complex and hard to comprehend, and the failure by those presenting the data to adequately communicate important information may result in undesirable or unpredictable decisions.

The purpose of this book is to provide a variety of skills and techniques for analysts in their presentation of statistical materials. It establishes a clear framework of alternative approaches to different target audiences so that complex and often confusing data, analyses, and arguments can be clearly and effectively communicated to decision makers.

Lawrence Witzling

Robert Greenstreet

Contents

Introduction

In many professional activities, it is necessary to present statistical data, quantitative analyses, and accompanying arguments to decision-making audiences. This is likely in engineering, urban planning, business, design, health administration, and similar professional fields in which the role of quantitative analysis is an integral part of successful practice. Size and scope of practice are no barrier to this need; the task of statistical communication will be necessary whether undertaken in a small office or a major public organization and may be carried out by one individual or by an entire agency.

The role of the analyst in this regard is to communicate quantitative information to the intended audience in a way that facilitates or accelerates the decision-making process. This process involves a transformation of raw data into understandable and palatable information via certain techniques, combining statistics with real-world constraints such as subjective values and judgments. The numbers can rarely speak for themselves, and failure to communicate adequately the meaning of the statistical data is likely to result in misunderstandings, continued ignorance of the relevant facts, and a failure to make necessary decisions either quickly or accurately.

Of course, education in quantitative analysis is very common, and dozens of courses are offered in high schools and universities across the country. Furthermore, the proliferation of the microcomputer has expanded the use of statistics, while computer graphics have increased the ability to create strong, visual images.

However, despite the plethora of courses and sophisticated technology, the analyst receives very little direction in developing the skills necessary to communicate basic data effectively. Each target audience has a different composition of interests and expertise, and may vary in size, attention span, and commitment to the issues being addressed. To be effective, the analyst must be able to alter the content and structure of the presentation to suit the specific needs of each target audience. This will require considerable skills, as the target audience may vary from a small group of in-house decision makers (such as a board of directors or working committee) to a large public gathering that, should its members voice their opinions, may secondarily affect decisions that ultimately have to be made by others. Furthermore, audience members may have widely ranging levels of expertise or skill in understanding quantitative analysis, which makes the analyst's role of communication even more complex.

The purpose of this book, therefore, is to provide a framework of skills and techniques for analysts with a basic grasp of computational methods, indicating alternative ways of organizing and presenting available data in a variety of circumstances to achieve the maximum impact. It is intended to develop within the reader an understanding of the importance of the presentation planning process, and to help the analyst to develop powers of critical thinking and creative judgment necessary to achieve consistently good presentations.

The presentation strategies in this book employ a variety of written and oral techniques, visual formats, and audience settings, and explore a wide range of alternatives for statistical preparation. For example, written presentations may take the form of a memo, summary report, full report, supplementary notes, or combinations of these formats. Similarly, oral presentations may be communicated with visual displays posted on walls, flipcharts, slides, computer screens, other audiovisual equipment, and combined techniques. In fact, the same statistical analyses may require multiple transformations to suit several different types of audiences and communications settings. One setting may be structured with continuous interaction or dialogue between the analyst and the audience, while another may be less conversational in character and more like a lecture or film presentation with little dialogue.

Although the analyst aspires to communicate effectively to multiple and varied audiences, there must be a simultaneous consideration of the practical constraints on the resources for presentation. Time, money, and expertise are likely to be limited, and not all relevant data may be readily available at the appropriate time. By developing a general attitude based on sound principles, the analyst can work efficiently within real constraints to produce the best possible results and achieve the ultimate goal: enabling decisions to be made based on the statistical material presented.

The contents of the book are laid out to provide the reader with a framework of information to facilitate the planning and development of future presentations. Chapter One looks at basic statistical descriptions and contains suggestions that are germane to many of the subsequent chapters. Chapters

Two and Three address, respectively, the problems of presenting statistical data for production (temporal information) and for geographic areas (spatial information). Chapters Four and Five examine more directly the use of statistics as components of formal decision-making methodologies such as benefit-cost analyses, cost-effectiveness, decision matrices, and decision trees or hierarchies. Chapter Six considers statistical techniques that are used to help plan, program, and/or implement time-limited projects, while Chapter Seven concludes with a series of basic strategies and outlines that can be used in conjunction with any of the information contained in the former chapters to ensure effective presentations.

The text is liberally supplemented with illustrations and practical examples of statistics in action, and concludes with a commentary of existing texts concerned with the more detailed explanations of statistical method. It is hoped that, in total, the book will provide a complete guide to the analyst wanting to develop an approach to present statistics effectively to a wide spectrum of potential decision makers.

1

Simple Statistical Descriptions

The most common forms of statistical presentation are the table, bar chart, and circle diagram, which often are used to present simple descriptive statistics such as frequencies and percentages. This chapter explores the possible uses and variants of these elementary statistical presentation techniques, although their widespread use in other types of presentation will be examined in subsequent chapters.

TABLES

Tables of numbers are one of the easiest and most flexible techniques to use for effective presentations. This section outlines general forms of tables and some basic rules for organizing their content.

Columns and Rows

The first decision when planning a tabular presentation will be how many columns and/or rows to create. The simplest table has just one column, which is, in effect, a list of numbers. If this is enough to make the intended point, the table should not be expanded simply to show more data.

The more common form of table, however, is a two-way table, where columns and rows intersect and each intersection becomes a cell. The smallest two-way table is a 2 x 2 matrix with two rows and two columns. Here, too, the numbers of rows and columns should not be expanded unless absolutely

TABLE A

AVERAGE DAILY ATTENDANCE DATA AND NET REVENUE PER SEAT
FOR MUDVILLE SPORTS STADIUM 1960- 1969.

		Old Stadium	New Stadium		
1960	ADA	7632	10320	ADA	1964
	RPS	1.14	.69	RPS	
1961	ADA	8510	15240	ADA	1965
	RPS	1.33	.86	RPS	
1962	ADA	4236	23868	ADA	1966
	RPS	1.89	1.35	RPS	
1963	ADA	6612	24112	ADA	1967
	RPS	1.34	1.71	RPS	
			33801	ADA	1968
			3.04	RPS	
			32001	ADA	1969
			2.88	RPS	

ADA = Average daily attendance
RPS = Revenue per seat

TABLE B

MUDVILLE SPORTS STADIUM: THE OLD VERSUS THE NEW.

		ATTENDANCE (daily average)	NET REVENUE (per seat)
OLD STADIUM	1960	7632	1.14
	1961	8510	1.33
	1962*	4236	.89
	1963	6612	1.34
NEW STADIUM	1964	10320	0.69
	1965	15240	0.86
	1966	23868	1.35
	1967	24112	1.71
	1968	33801	3.04
	1969	32001	2.88

* year following Casey strike-out

FIGURE 1.1 SWITCHING COLUMNS AND ROWS (*Explanation*: Table B is better because it is easier to compare the important numbers for attendance and revenue: both attendance and revenue can be seen in an uninterrupted sequence, the titles are easier to understand, and the graphic order is clearer.)

necessary. A two-way table occasionally may seem awkward if, for example, it has only two rows and a dozen or more columns, in which case, it may be worthwhile switching the columns for the rows in order to see which visual array is the most effective (Figure 1.1).

In addition to two-way tables, it is possible to construct three-way and even four-way tables by the use of subtitles, which create subsets of columns and/or rows. Such tables can be difficult to comprehend, but can contain a wealth of relevant information. Therefore, it is important to organize three-way and four-way tables in a graphically effective manner with a clear visual hierarchy that facilitates the viewer's ability to focus on individual numbers and at the same time to comprehend the overall picture being created (Figure 1.2). The section in this chapter entitled "Focusing the Audience" is particularly relevant in this regard.

Sequencing Columns and Rows

The juxtaposition of columns and rows can facilitate or hamper comparisons between data, although in some cases there may be little choice. For example, where each column represents successive numeric intervals (referred to as an interval *scale*), it would be confusing to sequence columns of income levels so that the first column was $0-$10,000, the second $50,000 and above, the third $20,001-$49,999, the fourth $10,001-$19,999, and so on. If the data must be arranged using a numeric sequence, the only reasonable choice is whether to begin at the low or high end of the scale. The same problem occurs when data are organized in a nonnumerical rank order (or *ordinal scale*). Here, too, the columns and rows have an explicit order, such as high, medium, and low.

There is , however, another more flexible way of sequencing data in a table, referred to as a *nominal scale*. While interval scales have a numerical order and ordinal scales have a clear nonnumerical ranking, nominal scales have no numerical or rank order, but are a distinct set of unique categories. For instance, the columns may each represent the income of persons in a different city or county. When columns or rows are based on a nominal scale, there is a significant amount of choice as to how they should be sequenced. They may perhaps be sequenced according to their relevance to the issues being discussed, where columns and rows can be organized to facilitate important comparisons between them and/or among key data items. One type of scale may be used in sequenced columns, whereas another type can be used for rows. If an interval scale is used for columns displaying income and an ordinal scale is used for rows displaying communities, there is little flexibility in shifting columns, but considerable choice in shifting rows. Therefore, the presentor can try out different arrangements to see which best fits the needs of the audience and the nature of the issues involved (Figure 1.3).

If it is desirable to focus attention on the relationship of two columns or rows of data, those two columns or rows should be adjacent. If the data structure does not allow this, graphic highlights may be used to emphasize significant columns, rows, and numbers. This will be dealt with in more detail in the last section of this chapter, entitled "Serial Comparisons," and in Chapter Seven.

Modifying Columns and Rows

The three types of scale — interval, ordinal, and nominal — can be modified in different ways. It may be advantageous to change the type of scale used to organize the columns and rows by modifying the subtitles. For example, if an

TABLE A   NUMBER OF PEOPLE RECEIVING IMMUNIZATIONS (IN 1,000)

		1970-1974			1975-1979		
		Children	Adults	Subtotals	Children	Adults	Subtotals
Program A	Schools	24	8	32	30	4	34
	Hospitals	12	25	37	2	35	37
	Clinics	18	55	73	22	60	88
			TOTAL	142		TOTAL	159

Program B	Schools	20	7	27	24	18	42
	Hospitals	18	33	51	9	21	30
	Clinics	22	42	64	25	52	77
			TOTAL	142		TOTAL	149

TABLE B   NUMBER OF PEOPLE RECEIVING IMMUNIZATIONS (IN 1000)

		CHILDREN			ADULTS		
		Program A	Program B	Subtotals	Program A	Program B	Subtotals
Schools	1970- 74	24	20	44	8	7	15
	1975-79	30	24	54	4	18	22
Hospitals	1970- 74	12	18	30	25	33	58
	1975-79	2	9	11	35	21	56
Clinics	1970- 74	18	22	40	55	42	97
	1975-79	22	25	47	60	52	112
Subtotals	1970- 74	54	60	114	88	82	170
	1975- 79	54	58	112	99	91	190

FIGURE 1.2 ORGANIZING A FOUR-WAY TABLE (FOUR SETS OF VARIABLES) (*Explanation*: These two tables each organize the same data to emphasize different variables and to facilitate different comparisons. Choosing one over the other depends on the audience and the problem at hand.)

analyst has an interval scale for age groups divided into six categories, each with specific age limits, it could be reorganized into a nominal scale of three categories labelled school-age children, elderly, and other (Figure 1.4). Similarly, a nominal scale identifying three population groups could be reorganized as an ordinal scale such as "primary target group." Both of these examples presume that reorganizing a table depends on the issues being analyzed and the presentor's understanding of the audience.

In general, however, there is a tendency among analysts to present data at the most complete level of detail. If, for instance, data on income were

TABLE A   FACTORS INFLUENCING COSTS OF EDUCATION

	Student Enrollment (in 1,000)	Number of Teachers	Annual Cost per Student
City A	13.5	500	$2050
City B	20.0	800	$1568
City C	90.0	3000	$2500
Average	41.2	1433	$2039

TABLE B   COMPARISON OF EDUCATIONAL COSTS BY CITY

	Annual Cost per Student	Student Enrollment (in 1,000)	Number of Teachers
City A	$2050	13.5	500
City B	$1568	20.0	800
City C	$2500	90.0	3000

TABLE C   PATTERNS OF TEACHER EMPLOYMENT

	Number of Teachers	Student Enrollment (in 1,000)	Annual Cost per Student
City A	500	13.5	$2050
City B	800	20.0	$1568
City C	3000	90.0	$2500

FIGURE 1.3 RESEQUENCING COLUMNS OR ROWS **(Explanation:** Each table sequences the three columns in a different way for a different purpose as is implied by the title of the table.)

available in a numeric or interval scale with ten or more categories, it is possible (although not at all usual in current practice) to transform the data effectively in a number of ways. The data can be collapsed into fewer catagories (such as three income ranges), or a new ordinal scale can be created with high-, middle- and low-income categories. Even a nominal scale can be developed with a target income group and a category of other (Figure 1.5). It should be remembered that, for maximum audience impact, the type of scale, and number of columns and rows and their titles should fit the needs of the audience and the argument and not the values of the analyst.

Parallel Construction of Tables

The same columns or rows occasionally reappear in several tables, as in the case of a series of tables, each looking at a different variable related to the same population groups. The tables then can be constructed in a parallel manner. If the population groups are the rows in the first table, they should be the rows in the subsequent tables. Similarly, the sequence of rows or columns that are repeated in several tables should remain constant.

When one creates parallel tables, a particular row or column does not need to appear continually or alternatively to have any data associated with it in a particular table. In such cases, the rows at the bottom of the table (or columns near the right side) may be located so that their absence does not disrupt the sequencing of rows or columns in the remaining tables (Figure 1.6). The alternative is to include the same rows or columns in all tables, just to create a sense of continuity. If the rows or columns are empty or not applicable, they can be labelled accordingly and deemphasized graphically.

TABLE A   POPULATION DISTRIBUTION (IN 1000) BY AGE FOR FOUR
COUNTIES IN THE SMSA.

Age (in years)

	0- 4	5- 17	18- 24	25- 44	45- 64	65 & over	Total
County A	6	13	26	40	35	20	140
County B	8	22	44	55	50	32	211
County C	20	58	88	125	110	98	499
County D	4	12	19	22	25	12	94
Totals	38	105	177	242	220	162	944

TABLE B   POPULATION OF CHILDREN AND ELDERLY IN METROPOLITAN AREA

	School-Age Children (0-17)	Elderly (65 and older)	Other (18-64)
County A	19	20	101
County B	30	32	149
County C	78	98	323
County D	16	12	66
Totals	143	162	639

FIGURE 1.4   COLLAPSING COLUMNS   (**Explanation**: Changing the data categories of Table A from an interval scale to a nominal scale creates Table B, which summarizes the data into new, more recognizable categories for an audience interested only in some of the information.)

Cells

Each box in a table is a cell, and each cell can contain more than one number. It is common for each cell in computerized cross-tabulations to contain up to four numbers, which are: the raw number or frequency; the column percentage (the frequency or number shown in the cell, divided by the column total); a row percentage (the same frequency divided by the row total); and a combined row and column percentage (the frequency divided by the grand total of all the cells). Such tables are easily understood by persons who prepare and review them regularly, but may be overly complex and confusing to an uninitiated audience (Figure 1.7).

To avoid this situation, the table should be designed so that each cell contains only a few numbers, effectively labelled and visually organized. Row and column percentages are the most common method, although it is possible to create other types of cells with multiple entries. Cells might contain data for the same variable at two different time periods, or for two different population groups (Figure 1.8). Each cell might contain a range, such as a *high estimate* and a *low estimate*. The choice here is whether to place multiple entries in each cell, to create separate columns or rows, or even to display separate tables for second or third sets of numbers. Multiple entries in each cell generally should be used only if the entries are linked mathematically (such as row numbers and percentages), or the audience is required to focus continually on comparisons between the multiple entries. For example, if the audience needs to see the same variable at two different time periods, multiple entries in one cell are appropriate. On the other hand, if the analyst wants to provide a comprehensive breakdown of the first time period and, after that has been assimilated, requires audience examination of the second time period, it would be more effective to use separate sets of columns, rows, or tables for each time period (Figure 1.8).

Focusing the Audience

The analyst wants the audience to read the statistical display in the order intended, but there are no conventional rules or habits, as in ordinary reading, that can predict the audience's focusing behavior.

An audience presumably will look first at the title and some of the subtitles if the words are visually prominent and reasonably simple. If the labels are relevant to the presentor's argument and the audience's interest, there is a good chance that the audience will begin to read and try to comprehend *some* of the numbers. How many numbers will an audience examine and try to comprehend? Obviously, each individual is different, and some will look at everything, others at just a few numbers, and some at nothing. It probably is safe to assume that most of the audience will look at three or four numbers, and possibly at as many as ten or twelve. However, even twelve numbers do not make a very large table — merely a 3 x 4 matrix.

Given the assumption of limited audience attention, there are various strategies that can be used to present larger tables of statistics more effectively. One way to focus audience attention may be to reduce the size of the table by combining irrelevant rows and columns into fewer categories (such as grouping them under the subtitles *other* or *not applicable*). Alternatively, titles and subtitles can be changed such that irrelevant data do not appear. Once the rows

TABLE A   INCOME DISTRIBUTION BY COUNTY IN THE SMSA   (POPULATION IN 1,000)

	Under $10,000	$10,001-$20,000	$20,001-$40,000	$40,001-$70,000	Over $70,000	Total
County A	3	6	25	29	5	68
County B	5	8	35	24	4	76
County C	7	20	45	55	10	137
County D	1	2	7	8	2	20
Totals	16	36	112	116	21	301

TABLE B   POPULATION (IN 1000) FOR INCOME CATEGORIES FOR METROPOLITAN AREA

	Low to Moderate Income	Middle Income	Upper Income	Totals
County A	9	54	5	68
County B	13	59	4	76
County C	27	100	10	137
County D	3	15	2	20
Totals	52	228	21	301

TABLE C   TARGET INCOME GROUP (IN 1000) IN CITY VERSUS SUBURBS

	Target Group (under $20,000)	Other	Totals
City (County C)	27 (52%)	110 (44%)	137 (46%)
Suburbs (Counties A, B, D)	25 (48%)	139 (56%)	164 (54%)
Totals	52 (100%)	249 (100%)	301 (100%)

Figure 1.5 Collapsing Columns and Rows (**Explanation**: The progression from Table A to Table B to Table C shows how data categories are collapsed to focus on concepts rather than statistics.

and columns are reduced to a manageable number, they should be organized fromleft to right and from top to bottom so that they correspond effectively to accompanying verbal arguments. It is important to establish which issue will be used at the beginning of an argument so that the first set of columns, rows, or cells can be made relevant to that issue.

There are, however, occasions when large tables must be used. Collapsing rows and columns even a little may make some members of certain audiences think their individual concerns are being ignored. Similarly, if a report contains too many tables, it may be wiser to combine statistics into fewer, albeit larger, tables. In other situations, it may be necessary to use the same sequence of columns or rows for several tables, thereby making it difficult to collapse the table matrix.

When large statistical tables are needed, the presentor may have to use certain graphic techniques to focus attention on special columns, rows, or cells. This can be accomplished with simple graphic devices, such as circling the

TABLE A   OPTIONS FOR RESIDENTIAL DEVELOPMENT (# UNITS BUILT)

Expected # Sales	OPTION A	OPTION B
1st year	35	45
2nd year	100	80
Total	135	125

TABLE B   MANAGEMENT COSTS FOR RESIDENTIAL DEVELOPMENT OPTIONS

	OPTION A	OPTION B	OPTION B.1
1st year	$175,000	$175,000	$125,000
2nd year	$225,000	$175,000	$200,000
Total	$400,000	$350,000	$325,000

TABLE C   OPTION B : EXPECTED SALES BY BUYER GROUP

	Families	Singles	Retirees	Total
1st year	20	20	5	45
2nd year	22	40	18	80
Years 3-5	0	3	12	15
Total	42	63	35	140

Figure 1.6 PARALLEL CONSTRUCTION OF TABLES (*Explanation:* When constructing a series of tables on the same subject, the columns and rows should retain a similar sequence wherever possible.)

number, using boldface type (for typeset materials or with some word processing systems) or using a separate color or tone, especially when presenting slides or large graphic displays on boards (Figure 1.9).

It is not possible, however, to focus attention on every number using even the most sophisticated graphic techniques. If too many numbers are given graphic emphasis, the overall effect may be as confusing as not giving emphasis at all. Occasionally, it may be necessary to bring all of the numbers in a large table into focus at different moments. This can be achieved by repeating the table several times, each time graphically emphasizing a different subset of numbers (Figure 1.10). This is particularly effective in slide presentations. It is somewhat more difficult to achieve when creating a series of large boards for an oral presentation, but can still be accomplished effec-

TABLE A   TARGET INCOME GROUP (IN 1000) IN CITY VERSUS SUBURBS

	Target Group	Other	Totals
City	27 20% / 52%	110 80% / 44%	137 100% / 46%
Suburbs	25 15% / 48%	139 85% / 56%	164 100% / 54%
Total	52 17% / 100%	249 83% / 100%	301 100% / 100%

TABLE B   TARGET INCOME (IN 1000) IN CITY VERSUS SUBURBS

	Target Group		Other		Totals	
City	27 / 52%	20%	110 / 44%	80%	137 / 46%	100%
Suburbs	25 / 48%	15%	139 / 56%	85%	164 / 54%	100%
Total	52 / 100%	17%	249 / 100%	83%	301 / 100%	100%

FIGURE 1.7 CELLS WITH PERCENTAGES (***Explanation***: Frequently, cells contain both raw data (integers) along with row and column percentages. Tables A and B indicate two ways to achieve this effectively, although Table C in Figure 1.5 shows how to present only one set of percentages.)

tively through the use of overlays and inexpensive photo reproduction techniques. In report form, it may not be necessary to reproduce the same table several times, but rather to introduce the entire table at the beginning of the report and then only repeat subsections of the table as needed. A more detailed discussion of graphic techniques can be found in Chapter Seven.

CIRCLES AND PIE DIAGRAMS

In addition to tables, the circle, or pie diagram, is a common form of presentation, where the size of each piece of the pie in a circle diagram is proportional to a statistical frequency. For example, if one of the numbers accounts for 25

TABLE A   TWO OPTIONS FOR RESIDENTIAL DEVELOPMENT (# UNITS BUILT)

	OPTION A		OPTION B	
	1st Year	2nd Year	1st Year	2nd Year
Management Cost (in $1,000)	175	225	175	175
Total Sales	35	100	45	80
Sales by Group: Families Singles Retirees	20 10 5	40 50 10	20 20 5	22 40 18

TABLE B   ANNUAL BREAKDOWN FOR RESIDENTIAL DEVELOPMENT OPTIONS

	First Year		Second Year	
	Option A	Option B	Option A	Option B
Management Cost (in $1,000)	175	175	225	175
Total Sales	35	45	100	80
Sales by Group: Families Singles Retirees	20 10 5	20 20 5	40 50 10	22 40 18

FIGURE 1.8 REORGANIZING CELLS (*Explanation*: By shifting the size, shape, and content of cells, different concepts can be emphasized. This example, based on data from Figure 1.6, shows how each table facilitates different visual comparisons, thereby affecting the audience in different ways.)

TABLE A   INCOME DISTRIBUTION BY COUNTY IN THE SMSA (POPULATION IN 1,000)

	Under $10,000	$10,001- $20,000	$20,001- $40,000	$40,001- $70,000	Over $70,000	Total
County A	3	6	25	29	5	68
County B	**5**	**8**	35	24	4	**76**
County C	**7**	**20**	45	55	10	**137**
County D	1	2	7	8	2	20
Total	**16**	**36**	112	116	21	301

TABLE B   INCOME DISTRIBUTION BY COUNTY IN THE SMSA (POPULATION IN 1,000)

	Under $10,000	$10,001- $20,000	$20,001- $40,000	$40,001- $70,000	Over $70,000	Total
County A	3	6	25	29	5	68
County B	5	8	35	24	4	76
County C	7	20	45	55	10	137
County D	1	2	7	8	2	20
Totals	16	36	112	116	21	301

FIGURE 1.9 EMPHASIZING SETS OF CELLS (**Explanation**: If a larger table must be used, the presenter can use different graphic techniques to focus attention on specific data. These examples are based on the same data as are shown in Figure 1.5. Each table emphasizes the same cells in a different way.)

percent of a total, that number is allocated 25 percent of the circle (that is, 90° out of 360°). The sum total of all of the statistics or slices remains 100 percent, or 360° of the circle. The graphic simplicity and visual immediacy of this technique makes it an excellent tool, although the graphic presentation must be handled skillfully, since the value of this technique rests solely on its visual clarity.

The number and size of the pie slices initially should be controlled, and smaller slices can be combined into larger categories to reduce visual clutter (Figure 1.11); there is no analytic or ethical value to giving every number or percentage its own slice, simply because it has been calculated.

Important numbers to be presented to the audience should be identified and, if particular numbers are not relevant, should be combined and included in a slice labelled "other." It may also be appropriate to leave them in a blank space. Another option may be to redefine and retitle the set of numbers being examined so that only a relevant subset of statistics are included in the circle diagram. This subset can be recalibrated so that it totals 100 percent and completes the 360° of the circle (Figure 1.11).

After the number and size of the pie slices have been determined, their sequence in the diagram must be established. Most viewers probably read the largest slice first and/or the slice located in the upper left. Subsequent slices are probably read in clockwise and/or descending order of size. In general, the largest slice (i.e., the largest percentage) should be located in the upper left or left center, and then subsequent percentages should be presented in descending order in a clockwise fashion.

There is, however, one important exception to this rule. If the verbal presentation implies a specific sequence to reviewing the statistics, that sequence should be used to generate the order of the pie slices.

An especially effective variation of the circle diagram illustrates one or more salient slices pulled out from the circle. Each of these slices can be subdivided further. This allows the presentor to show two levels of statistics. The slice that is pulled away can be shown relative to the rest of the circle, while the components of that slice can show a second, more detailed, level of statistics (Figure 1.12).

PARALLEL BARS

The use of parallel bars to illustrate relative frequencies is as common as the use of pie or circle diagrams, and is equally as simple, where the height of each bar is proportional to the number it represents. Parallel bars typically are used to show a series of comparisons of up to four sets of numbers. That is, there are multiple sets of numbers, with each set showing a different comparison. Most common are comparisons over time where, for example, the size of two groups are compared at year 1, year 2, year 3, and so on.

Parallel bars are most useful when two kinds of comparisons must be presented simultaneously, such as a comparison of the incomes of different groups of people, and incomes at different times or different places. It is first necessary to decide how to group the bars, and usually a set of bars can be assembled in at least two ways, because there are at least two issues or variables being presented. In the example of income levels, the data could be presented either with one set of bars for each year or with one set of bars for each population group (Figure 1.13).

It is usual to select the variable or issue with the fewest subcategories as the basis for the adjacent bars and the variable with the greater number of categories as the basis for separating the sets of bars. For instance, in presenting the income levels for two population groups (A and B) over five years (1 to 5), five sets of paired bars would be displayed as follows: $A1$ with $B1$, $A2$ with $B2$, $A3$ with $B3$ and so on. If the income levels of five groups (A through E) at two time periods (1 and 2) were being compared, the pairs of bars would probably be $A1$ with $A2$, $B1$ with $B2$, up to $E1$ with $E2$. There are, of course, exceptions to this rule. If, for example, each variable or issue has three to five categories, there is some ambivalence as to the most legible organization for the data.

A more important question governing the organization of parallel bars concerns any verbal presentation that accompanies the data. The bars must be organized to facilitate the comparisons critical to the issues, and the most prominent feature of such diagrams is the difference in the height of adjacent bars. Consequently, changes in height differences from one set of bars to the next should govern the visual organization of the diagram. In the example

TABLE B   POPULATION OF CHILDREN AND ELDERLY IN METROPOLITAN AREA

	School-Age Children (0-17)	Elderly (65 and older)	Other (18-64)
County A	19	20	101
County B	30	32	149
County C	78	98	323
County D	16	12	66
Totals	143	162	639

TABLE B   POPULATION OF CHILDREN AND ELDERLY IN METROPOLITAN AREA

	School-Age Children (0-17)	Elderly (65 and older)	Other (18-64)
County A	19	20	101
County B	30	32	149
County C	78	98	323
County D	16	12	66
Totals	143	162	639

TABLE B   POPULATION OF CHILDREN AND ELDERLY IN METROPOLITAN AREA

	School-Age Children (0-17)	Elderly (65 and older)	Other (18-64)
County A	19	20	101
County B	30	32	149
County C	78	98	323
County D	16	12	66
Totals	143	162	639

FIGURE 1.10 SEQUENTIAL EMPHASIS OF CELLS (*Explanation*: In some presentations, the same table may be repeated to emphasize different concepts. This illustration, based on the data from Figure 1.4, presumes that the presentor is focusing the audience upon a sequence of issues.)

DIAGRAM A   PERCENTAGE OF POPULATION BY INCOME RANGE IN SMSA

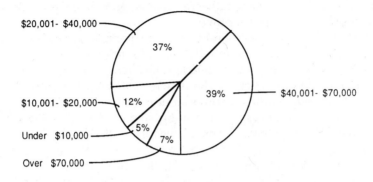

DIAGRAM B   POPULATION BY INCOME GROUP IN METROPOLITAN AREA

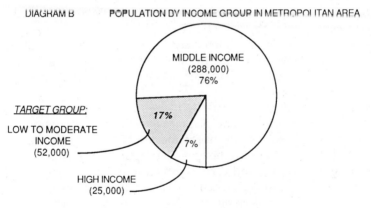

DIAGRAM C   TARGET INCOME GROUP (IN 1000) BY COUNTY

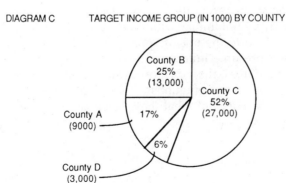

FIGURE 1.11   COLLAPSING CIRCLE DIAGRAMS   (**Explanation**: Data in circle diagrams can be recombined in different ways to emphasize various issues. These examples are based on the same data that are used in Figure 1.5.)

previously used, the goal may be to focus attention on the differences between groups. The bars should then be paired for population groups (not time periods). Conversely, time periods rather than population groups should be combined into one set of bars (rather than paired population groups) if the goal is to focus on differences over time.

Sometimes, it may be useful to combine nonrelevant data into a bar or category labelled "other," or to redefine and retitle the bar diagram so that irrelevant data is simply eliminated. When reorganizing and combining parallel bars, it is more effective to limit each set of bars to two or three as visual comparisons become difficult with four or more bars in each set.

Another important aspect of creating bar diagrams lies in the selection of a vertical scale and/or the use of break lines or cuts to truncate the length of all bars. The audience is most interested in the height differences among bars. However, such height differences are looked at in relation to their entire length. The relative length of the bars can be changed dramatically by selecting different starting points for the vertical scale and the use of break lines. For example, by selecting a low number for a starting point, there will be a visually smaller difference in height between two bars. The audience will perceive it as a smaller relative difference, which may or may not be the goal of the presentation. Conversely, truncating the height of parallel bars by using break lines or a high starting level may exaggerate a difference beyond the intent of the presentation (Figure 1.14).

The last issue concerning parallel bars is more technical. A graphic device—similar to the use of bar diagrams—called a histogram is often described in statistical texts. Briefly, in a histogram, bars are of unequal widths, and the area of the bar—not the height—is proportional to the number or statistic. Both the horizontal and vertical dimensions are interval scales with numeric sequence (Figure 1.15), whereas in a bar diagram, only one dimension—usually the vertical—is an interval scale. Histograms can be useful, and often may seem to be the most accurate way to reflect the data. However, the general public is relatively unfamiliar with their meaning, and the use of a histogram may easily lead to the misinterpretation of data if the audience members automatically presume, in accord with their past experience, that the height of a bar and not the area is the indicator of value. Histograms should not be used for general audience presentations without a thorough explanation and clear examples at the outset of what they mean and how they must be interpreted.

PILING DATA

A computation not often used for presentations is a cumulative frequency diagram, where the frequency of the number in the first category is added to the second, the third is added to that subtotal, and so on. Although it is not a

DIAGRAM A   POPULATION BY INCOME GROUP IN METROPOLITAN AREA

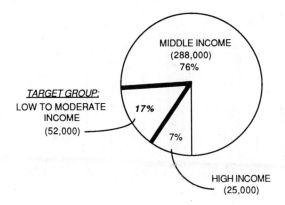

DIAGRAM B   TARGET INCOME GROUP (IN 1000) FOR METROPOLITAN AREA

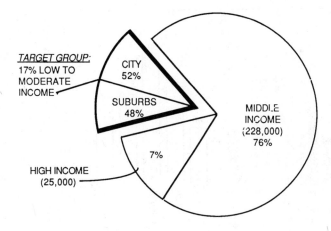

FIGURE 1.12 CIRCLES AND PIE SLICES (**Explanation**: By "pulling out" a pie segment of a circle diagram, the presentor can show two levels of data. This example is derived from Figures 1.5 and 1.11.)

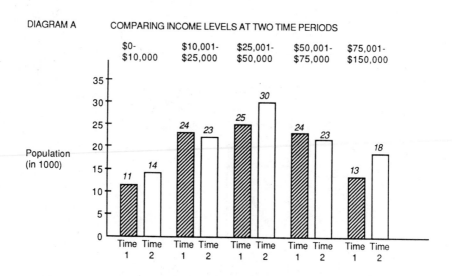

DIAGRAM A COMPARING INCOME LEVELS AT TWO TIME PERIODS

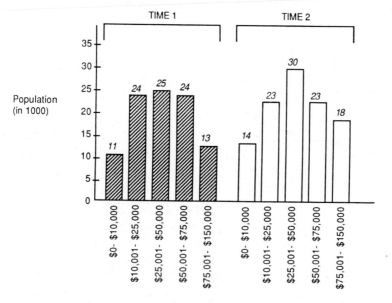

DIAGRAM B DISTRIBUTION OF INCOME AT TWO TIME PERIODS

FIGURE 1.13 COMPOSING PARALLEL BARS (***Explanation***: Each diagram presents the same data. However, the best grouping of the bars depends on which comparison is most important for the analytic argument.)

common procedure, it can be effective if the presentation involves a serial buildup of statistics.

If an analysis is concerned with how many persons in each of several groups exhibit a specific characteristic, such as a health problem or political opinion, it may be useful to show the data in additive form. In addition to showing one total at the end of a table or diagram, the presentation can show accumulations of data analogous to a running subtotal.

This typically is done by starting with the largest number (or frequency) and adding successively smaller numbers. These data can be displayed with a table, bar graph, or a more technical diagram called a cumulative frequency polygon (Figure 1.16). The latter is more difficult for general audiences to comprehend and, like a histogram, should be used only with full explanation and clear examples of how it is computed and interpreted.

SETS

Set, or Venn, diagrams are infrequently used, but are also effective presentation tools. In their simplest form, there usually are two to four overlapping circles, each representing a different set or category of data. An intersection of two circles includes data that fall into both categories or sets. For example, if circle A represents *the elderly population* and circle B represents the population in *need of additional health care* (Figure 1.17), the intersection contains the number and/or percent of people who are elderly *and* in need of health care. The remainder of circle *A* shows the number of elderly who do not need additional health care, while the remainder of circle *B* shows the number of persons who need additional health care but who are not elderly. The area outside both circles shows the number of persons who are neither elderly nor in need of more health care.

The same data could be shown in a simple 2 x 2 table. The advantage of the set diagram is that it focuses audience attention on specific numbers and issues. In this example, it is the intersection of the sets that presumably represents a *target* population. The other statistics in the circles but outside the intersection appear to be less relevant. Statistics describing values completely outside both circles take on even less visual relevance.

Another potential advantage of using set diagrams lies in the size of the circles, which can, in some cases, be made proportional to the numeric values in the sets. This has some of the advantages of pie or circle diagrams. However, it should be remembered that the set diagram is intended only to show logical relationships among data, and using the size of the circle to represent the relative size of numbers is an unconventional and invalid procedure.

Although circles are commonly used to enscribe sets, other shapes, such as rectangles and combinations of shapes, can also be effective. For example, one large rectangle could represent persons needing additional health care. A

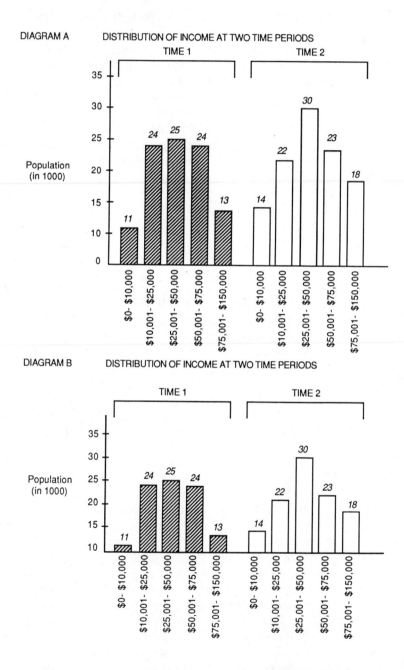

FIGURE 1.14 PARALLEL BARS AND VERTICAL SCALE (**Explanation:** Altering the vertical scale can change the image and impact of a diagram. These two examples portray the same data shown in Diagram B of Figure 1.13.)

DIAGRAM A   DISTRIBUTION OF INCOME AT TWO TIME PERIODS (BY POPULATION IN 1,000)

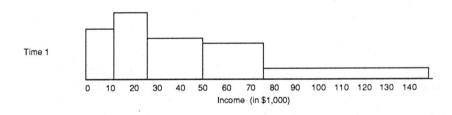

DIAGRAM B   DISTRIBUTION OF INCOME AT TWO TIME PERIODS (BY POPULATION IN 1,000)

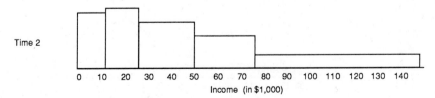

FIGURE 1.15 HISTOGRAMS VERSUS PARALLEL BARS (*Explanation:* In a histogram, there is no vertical scale, since the area inside each bar or segment of the diagram is proportional to the frequency of the number. These two histograms show the same data as are shown in Figure 1.13, Diagram B. Notice that the width of each bar varies with the size of the income level being represented. The selection of the size of the areal unit by which to calculate the areas is at the discretion of the analyst.)

DIAGRAM A   CUMULATIVE FREQUENCY OF INCOMES AT TWO TIME PERIODS

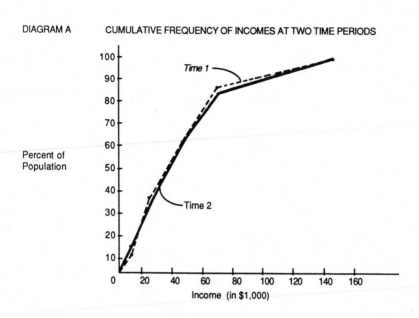

DIAGRAM B   CUMULATIVE FREQUENCY OF INCOMES AT TWO TIME PERIODS

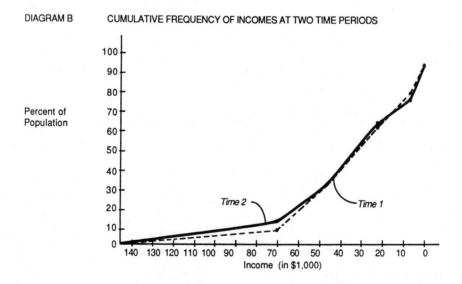

FIGURE 1.16 FREQUENCY POLYGON-COMPILING DATA (*Explanation*: Occasionally, it is useful to show how a series of numbers add together to form a 100 percent total. These two diagrams show how data from Figure 1.13 can be aggregated in two different ways.)

series of overlapping shapes could then represent several different target population groups (Figure 1.18).

When more than three sets must be overlapped, the diagram can become very complex, and alternative shapes, sizes, and graphic symbols should be explored to decide which best represents the major presentation issues. In presentation, various graphic techniques can be used to ensure the proper visual emphasis to different data sets, subsets, and intersections (Figure 1.18).

SERIAL COMPARISONS

Many presentations, as noted previously, require the display of several tables or diagrams organized in parallel structure and/or linked with graphic

DIAGRAM A   ELDERLY POPULATION AND HEALTH CARE NEEDS

Population (in 1,000)

	Elderly	Other	Totals
Special Health Care Needs	30 (38%) (75%)	50 (62%) (24%)	80 (100%) (32%)
No Special Need	10 (16%) (25%)	160 (94%) (76%)	170 (100%) (68%)
	40 (16%) (100%)	210 (84%) (100%)	250 (100%) (100%)

DIAGRAM B   THE ELDERLY AND HEALTH CARE NEEDS (POPULATION IN 1,000)

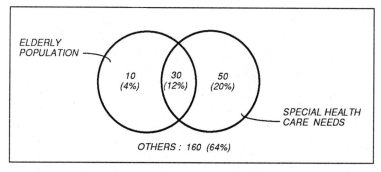

FIGURE 1.17 SET DIAGRAMS VERSUS TABLES (**Explanation**: In some circumstances it is more compelling to illustrate data by using set diagrams rather than tables. The data for the table in Diagram A are the basis for the sets shown in Diagram B.)

techniques. This can be taken one step further by linking together tables with a verbal argument. A logical chain of small tables or diagrams is an effective technique for leading an audience through complex statistics that focus on a single issue.

For example, an argument supporting an expanded program for creating new jobs may have to be developed, and statistics on several geographic areas, types of workers for different industries, and different programs are available. However, the focus needs to be on one program in one target area for one type of new job opportunity, so one large table could be displayed emphasizing the most significant column, row, and intersection (Figure 1.19). When such a table

DIAGRAM A   SPECIAL HEALTH CARE NEEDS FOR DIFFERENT SEGMENTS OF THE POPULATION (IN 1,000)

		Special Health Care Needs	No Special Needs	Totals
ELDERLY	w/o handicap	25 (10%)	10 (4%)	35 (14%)
	w/ handicap	5 (2%)	0 (0%)	5 (2%)
CHILDREN	w/o handicap	18 (7%)	40 (16%)	58 (23%)
	w/ handicap	2 (1%)	0 (0%)	2 (1%)
OTHER PERSONS	w/o handicap	26 (10%)	118 (47%)	144 (57%)
	w/ handicap	4 (2%)	2 (1%)	6 (3%)
	Totals	80 (32%)	170 (68%)	250 (100%)

DIAGRAM B   SPECIAL HEALTH CARE NEEDS FOR DIFFERENT SEGMENTS OF THE POPULATION (IN 1,000)

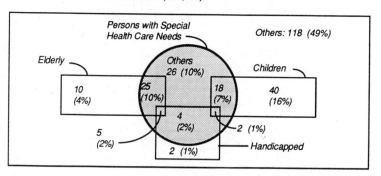

FIGURE 1.18 SET DIAGRAM VARIATIONS (*Explanation*: These two diagrams show how data from a table can be transformed into a multiple-set diagram by using different shapes.)

includes too many numbers, there is a clear risk that the audience will not focus on the key statistics or will be distracted by irrelevant or misleading statistics. They may reach conclusions prematurely before they have heard or read the accompanying verbal argument.

The alternative is to create a series of tables that ultimately lead the audience, in the last table, to the target issue. This technique helps avoid pitfalls by interweaving statistics with verbal arguments one step at a time, and by reducing the total volume of numbers to be presented. The first table in the series might cross-tabulate types of unemployed workers by different areas, and be used to identify the target area. The next table would cross-tabulate types of job opportunities, but only in the target area, highlighting the kinds of jobs that are the focus of the program. The last table would cross-tabulate different programs related to targeted job opportunities in the target area. The sequence of tables helps the audience focus on the sequence of issues and the logic being used (Figure 1.19). The three tables together used far fewer cells than the single comprehensive table. Of course, if the logic of the sequence is flawed, there is no way for the analyst to retreat behind a protective mesh of numbers. On the other hand, if the argument is sound, this technique will make the point more forcefully, avoiding the problems inherent in larger tables and providing a clear, comprehensible presentation that unfolds in a simple sequence of ideas.

TABLE A   JOB PROGRAM NEEDS AND IMPACTS

		Community A		Community B		Community C	
		Skilled Workers	Unskilled Workers	Skilled Workers	Unskilled Workers	Skilled Workers	Unskilled Workers
	Unemployed	10,000	3,000	15,000	10,000	13,000	37,000
	% of Population	7%	5%	8%	7%	10%	14%
	Potential for new job opportunities (next 3 years)	400	100	600	250	900	1700

EXPECTED JOB CREATION	Firms with 100 employees or fewer:
	     Program A
	     Program B
		100
		 200
			50
			 50
				200
				 200
					50
					 50
						200
						 100
							700
							 400

	Firms with over 100 employees:
	     Program C
	     Program D
		50
		 0
			0
			 0
				0
				 0
					0
					 0
						100
						 50
							0
							 100

FIGURE 1.19 TABLE SERIES (*Explanation*: In some cases, it may be more effective to present data in a series of tables, helping the presentor to direct the audience to a conclusion. Table A, for example, can be recast as a series of Tables (B, C, and D), which can emphasize relevant data in a variety of ways.)

TABLE B   UNEMPLOYMENT BY TARGET AREA

<table>
<tr><td rowspan="4">UNEMPLOYMENT</td><td colspan="2"></td><td>Community A</td><td>Community B</td><td>Community C</td></tr>
<tr><td rowspan="2">Skilled Labor</td><td>Number of People</td><td>10,000</td><td>15,000</td><td>13,000</td></tr>
<tr><td>% of Population</td><td>7%</td><td>8%</td><td>10%</td></tr>
<tr><td rowspan="2">Unskilled Labor</td><td>Number of People</td><td>3,000</td><td>10,000</td><td>37,000</td></tr>
</table>

		Community A	Community B	Community C
Skilled Labor	Number of People	10,000	15,000	13,000
Skilled Labor	% of Population	7%	8%	10%
Unskilled Labor	Number of People	3,000	10,000	37,000
Unskilled Labor	% of Population	5%	7%	14%

TABLE C   POTENTIAL FOR NEW JOB OPPORTUNITIES IN COMMUNITY C (NEXT THREE YEARS)

	Firms with 100 or fewer employees	Firms with over 100 employees	Totals
Skilled Labor	600	300	900
Unskilled Labor	1200	500	1700
Totals	1800	800	2600

TABLE D   EXPECTED PROGRAM IMPACT IN COMMUNITY C

	Program A	Program B
# of Jobs created in firms with less than 100 employees	900	500

	Program C	Program D
# of Jobs created in firms with over 100 employees	100	150

2

Predictions:
Temporal Statistics

Any prediction of future events is a guess or at best an educated guess, and most audiences are aware of this. This can be said of descriptions that infer present and past events, although most audiences seem to find inferences of past and present occurrences far less controversial. Consequently, when making a prediction it is almost always advisable to present a caveat—a warning as to the potential misinterpretation or misuse of the data and the conclusion. For example, when making predictions, the analyst may use qualifying statements such as: *"I don't have a crystal ball, but in my opinion...."* Such a simple qualification for a prediction indicates to the audience that the presentor has confidence in his or her prediction, but is aware of the potential for inaccuracy.

Caveats for predictions can be complex and require detailed explanation, as the presentor has to qualify predictions without either overstating or understating the case. An effective way to begin is by stating the degree of confidence the presentor has in the prediction and then outlining the rationale for such confidence. The rationale might include statements such as:

The data supporting this prediction are very reliable and carefully measured.

There are many predictions similar to this one, made by other individuals or organizations involved in similar investigations.

This prediction is for a relatively short term. There are few potential changes in circumstances which would cast doubt on its validity.

Several experts who were asked about this prediction agreed that it was reasonable.

Other statements used to qualify and support predictions are noted throughout this chapter and should be used carefully, as audience reaction to a prediction can be controversial. Unless the analyst is confident that an audience is unlikely to challenge a prediction, some qualification at the outset of the presentation should be included. The degree and severity of qualifications may vary according to the nature of the issue and the character of the audience. Without an appropriate caveat, however, an entire argument can be destroyed or grossly misinterpreted because the presenter has failed to establish the logical basis or ground rules for examining the prediction.

TIME LINES, CURVES, AND BRANCHES

The most common form of statistical prediction is a linear or curvilinear trajectory, which is more difficult to compute than simple descriptive statistics because it requires a statistical inference. The techniques often used are linear regression, multiple regression, curvilinear regression, and/or curve fitting. It is the purpose of this section to outline their potential uses but not to explain computational procedures, which can be found in a number of other statistics texts (see references in Appendix).

In a trajectory prediction, the usual presentation involves a graph with two axes, with the horizontal axis representing the variable, or issue, being predicted. The time periods usually begin at or near the left side, with the earliest point at which data are available. For each time period, there should be a point on the graph clearly denoting the numeric value of the associated variable. There also should be a clear denotation on the horizontal axis of the current time period, the future time periods for which predictions are being made and the numeric value of those predictions.

Picking a Projection Line

The first presentation question concerns a computational issue: What line or curve best fits the data? The analyst usually tries to fit a few different lines or curves to the data and selects the one that has the lowest degree of statistical error. The concept of *statistical error* is explained in detail in most textbooks for statistical computation and is expanded briefly in other sections of this chapter.

From a broader analytic perspective, however, there may be reasons for selecting a line or curve that has a greater statistical error but that fits other, equally valid criteria for making predictions. For example, where a prediction for income levels or health status is made with a straight-line projection and has the lowest statistical error, it may contradict well-established, validated theories as to how income levels or health status are likely to change. Furthermore, experts with reputations for sound judgment might indicate that special circumstances, absent in past time periods, are likely to occur and therefore to alter the path of future events.

The presentor should not automatically have faith in a prediction because it has the lowest statistical error. There may be other more valid reasons to select computational formulas that imply greater statistical error but that are more germane to other knowledge about the issue under discussion.

Connecting the Dots

Regardless of which line or curve is computed, the analyst has to choose a suitable presentation technique. A common method is to *connect the dots* by drawing a line from each data point to the next successive point in time, up to the present. This produces a sawtooth diagram with which most audiences are familiar. One problem with this technique is that points along the connecting lines often are interpreted—quite incorrectly—as implying interim numeric values for the variables or issue being examined. For instance, if a line is drawn between an income level in 1960 of $10,000 and an income level of $20,000 in 1970, a line connecting these two plotted points implies that the income level in 1965, the midpoint, would have been $15,000, which may not have been true (Figure 2.1). It is an inference, not a measurement. Such inferences may seem inconsequential if, for example, the time span between the points is relatively short compared with the overall time span shown in the diagram.

Nevertheless, this may be misleading. Such inferences should be either clearly noted, avoided altogether by not connecting the plotted points with lines or handled by connecting the points with dashed or dotted lines, to depict some measure of uncertainty.

Swaths and Branches

An alternative to connecting the dots is to draw a single projection line and let the dots float around it. This technique results from the computational process of determining the equation for the straight line or curve that comes closest to the existing data points.

The computation technique usually is linear regression for straight lines or a form of multiple regression for curved lines. The equation for the line or curve is supposed to minimize the level of statistical error, a concept that bears some explanation. The expression *statistical error* refers to formulas that measure the distance between the projection line and the plotted points. It does not mean that the projection is incorrect, but it suggests that the lower the statistical error, the better the fit between the projection line and the data.

DIAGRAM A   INCOME OVER TIME

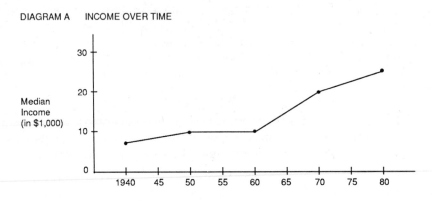

DIAGRAM B   INCOME OVER TIME

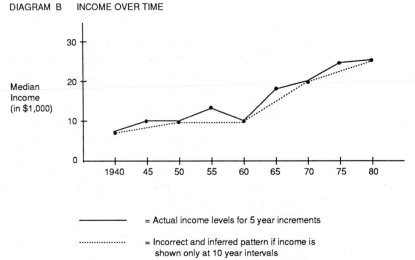

——————— = Actual income levels for 5 year increments

················· = Incorrect and inferred pattern if income is
                  shown only at 10 year intervals

FIGURE 2.1   CONNECTING THE DOTS   (***Explanation***: These two diagrams illustrate a simple problem—that connecting the points on a time-line diagram is potentially misleading if it infers intermediate values that have not been verified.)

    The projection line only approximates the value of the variable over time and is not an accurate measure of that value. This can be easily illustrated by showing the difference between the recorded value of the variable for a past point in time and the value indicated by the projection line for that same time. For instance, a projection line for income level will pass through previous time periods in which income has been recorded. The actual income level for a previous time period may be $25,000, but the projection line will pass through a different value (perhaps $20,000), indicating the size of the error for that time period (Figure 2.2).

It is important to indicate this degree of error so that the audience has a sense of the potential accuracy of the prediction. In another example, a projection line for income for past time periods has a typical error of $5000. It is reasonable for the audience to assume that there may be errors of about $5000 for predictions for future time periods. Too often, however, the audience is asked to believe that the predictions for future values of a variable are highly accurate and can be relied upon as a firm basis for decisions.

There are several ways to present projection lines to avoid such misinterpretations. The simplest method is to draw two lines rather than one that depict the edges of a swath, or wide time bar. The swath could include almost all of the existing or recorded data points for past time periods. The edges of the swath indicate the range of error implied by the mathematical equation for the line (Figure 2.2). As the swath moves into future time periods, the audience is not asked to believe that a specific value for a variable will occur, but rather that a range of values may occur.

A slight modification of this technique includes a third, center line in the swath that represents the midpoint of the range. In fact, the center is the single projection line that is normally drawn without indicating any statistical error functions.

A third, more complicated variation of this technique involves drawing a series of parallel projection lines with a visually dominant center line, and a series of parallel lines that become successively less dominant as they move away. This visually portrays the inference that the error for a prediction increases gradually as values move away from the center line.

Calculating the precise edge of a large swath or a series of parallel lines can be difficult. If there is insufficient time to calculate accurate computations, visual estimates of the edges or parallel lines may be appropriate. This option is not mathematically accurate, but does communicate the basic principle. It is relatively easy to achieve this kind of visual approximation for straight-line projections, but obviously more difficult with curves.

Another approach that appears similar, but which is, in fact, based on a different principle, involves drawing a single projection line up to the current time period and then drawing branched or splayed lines into the future time periods. This technique presumes that the projection line accurately represents past values of a variable, but that errors in predictions will increase as the projection line moves further into the future.

There is no precise way to compute branched projection lines, but they can communicate an important point to the audience by implying that decisions that require precise predictions become increasingly problematic as the predictions extend farther into the future.

Multiple Variables

Projection lines are sometimes based on multiple variables. In the case of a prediction of income for a community showing a wide degree of error, this becomes visually evident when the projection line is plotted. However, there may be reason to believe that income is influenced by education (or age, sex, race, or any other variable). In these situations, it is common to prepare a more

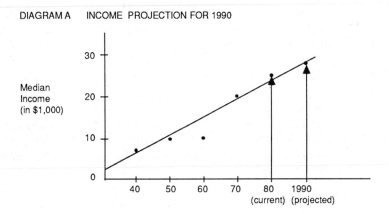

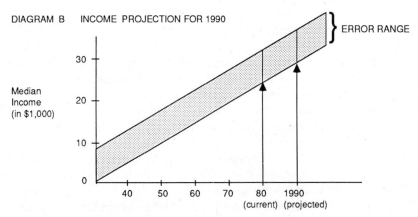

FIGURE 2.2 SHOWING THE "ERROR" *(Explanation*: An inferred projection line (typically based on linear regression or similar computations) always has a margin of error. Diagram A shows such a line without any indication of potential errors. Diagrams B, C, and D show ways of portraying such errors for a general audience. These diagrams are based on the same data as are shown in Figure 2.1.

complicated prediction based on multiple variables. Thus, a projection line for income for a group with the same level of education (or age or race) may have a smaller degree of error (Figure 2.3).

Predictions with multiple variables usually are based on multiple regression analysis or similar computational techniques. This practice is common in the social sciences and has the obvious merits of reducing the error associated with predictions. Most importantly, it allows for an apparently more reason-

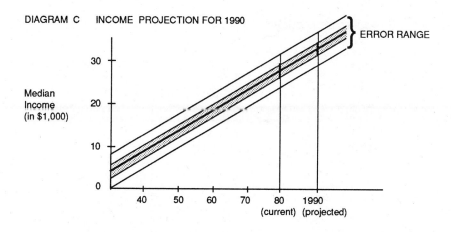

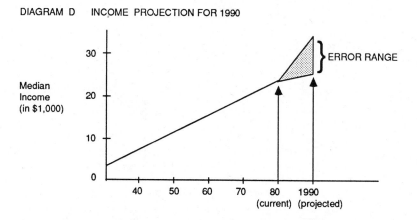

able approach to prediction, so if there appears to be a good reason to believe that two or more variables are linked together over time, this should be accounted for in the prediction.

Such prediction techniques are comprehensible to analysts who understand basic inferential statistics, but for general audiences, multivariate computational analyses appear far from simple, and create much greater presentation problems. A possible approach to prevent confusion is to begin with diagrams showing how each variable relates to the major, dependent variable. By using the preceding income projection example, one can develop a separate diagram or table for each of three "independent" variables that relate to

DIAGRAM A   AGGREGATE INCOME PREDICTION FOR ALL LEVELS OF EDUCATION

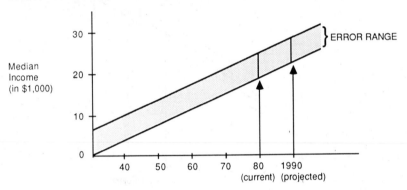

DIAGRAM B INCOME PREDICTION FOR PERSONS WITH LESS THAN HIGH SCHOOL EDUCATION

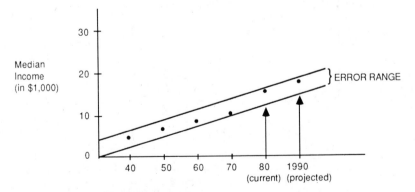

FIGURE 2.3 MULTIPLE PREDICTIONS (**Explanation**: In some cases, the analyst may want to show more than one prediction to portray the nature and size of the potential error or uncertainty. In this example, the data used in Diagram 2.2 could be separated into two diagrams, each with different degrees of error (Diagrams B and C).)

income, such as age, sex, and education. A fourth, more dominant, diagram would portray a projection of income *adjusted for age, sex, and education*. In this case, however, the vertical axis could not be labelled "income" because it would not be accurate; it would have to be labelled "adjusted income," and a full explanation must be offered to the audience.

Combined Variables: Indices

A similar approach to projections based on several variables is to invent a new variable, labelled index. Some indices are well known because of their frequent use in newspapers and on television, including the *consumer price index*, the *index of leading economic indicators*, and the *temperature humidity index*. Few people know how these indices are calculated, but many audiences accept them as meaningful measures of abstractions such as inflation, economic growth, and climatic comfort.

In the income example noted in the previous subsection, there might be a variable called an *income index*. The index would be some mathematical composite of several variables such as income, age, education, sex, and race. This new income index would not be expressed in dollar amounts, but would be more abstract and take on values—for example, from 1 to 100. The income

DIAGRAM C INCOME PREDICTIONS FOR PERSONS WITH HIGH SCHOOL EDUCATION

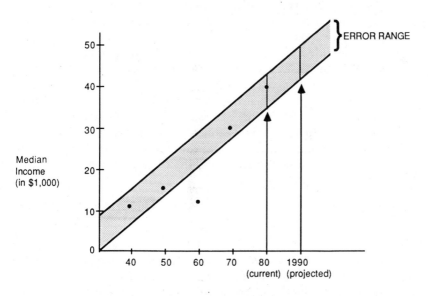

index then would be calculated as a single variable for past time periods and projected into the future.

The advantage of this technique is that it allows some flexibility to the presentor in compiling several numbers or several issues into one variable. The disadvantage is that the creation of a new, abstract measure creates the added burden of explaining the new concept. The audience must be convinced that a new variable such as an income index is meaningful and relevant to the accompanying verbal arguments or narratives.

Consequently, the validity or legitimacy of an index depends on verbal arguments and reasons, not on computational procedures. An index can be constructed in more than one way. The income index could be based on a multiple regression equation that minimizes the level of statistical error, or it could be based on the average income for ten sets of typical families or households. It may or may not be "adjusted for inflation."

An index can be made to measure relatively concrete concepts (such as income or prices) or more abstract notions. For example, if it was necessary to predict how many acres of land would be feasible for future development, a *feasibility index* might be constructed. There is no single phenomenon called feasibility that can be defined in the way age, income, or education can be measured. Feasibility is a concept with many components, some of which can be measured and combined mathematically into a new number indicative of the feasibility concept. In this instance, a feasibility index for developable land might be based on combining, at different time periods, the size of the home-buying population, forecasts for construction costs, proximity to major transportation routes, and income changes. The index can then be predicted as a projection line (Figure 2.4).

Projecting one index rather than several variables may imply a lower level of statistical error. However, creating an index has a more important function, even if it does not reduce statistical error, by focusing audience attention on the primary issue. If each variable related to feasibility (such as population, construction cost, income, transportation) were projected separately, the audience's attention might shift to multiple issues, fragmenting their attention and making the issues more difficult to comprehend. In this example, the feasibility index allows the presentor to focus attention primarily on the issue and then on the prediction.

THE S-CURVE

Sometimes a particular variable may appear to increase (or decrease) slowly, accelerate, and then level off. If such an observation is placed on a graph with a time line, it is shown as an S-shaped curve that is, in fact, a specific form of projection line. Computational formulas for S-curves are referred to as logistic, or Gompertz curves (see Appendix for references).

Equations for the S-shaped curves may not have the least amount of statistical error for a given data set, but can still be selected on the basis of whether or not they make common sense, given the presentor's and audience's understanding of all the issues at hand.

For example, past time periods may indicate that a slight upswing is beginning. An examination of alternative projection lines might show that a straight line is the best statistical fit, even though there is good reason to believe that a major change is at hand that would push the trend dramatically upward (Figure 2.5). In another instance, the data may show that there is a continuing, sharp rise in a particular variable, and the best statistical fit is a curve that is bending vertically at a steep angle. However, future expectations based on a long-term continuation of such an upward angle may appear unrealistic. In both of these cases, the use of an S-shaped curve based on theoretical, experiential, or otherwise nonstatistical issues may be a far more convincing and probable course of events.

The use of an S-shaped curve can seem arbitrary when it has a relatively higher degree of statistical error, but if there are sound reasons for suspecting that an S shaped curve makes sense, it should be presented along with the specific nonstatistical rationale for its use.

The S-shaped curve, like projection lines, can be presented in several ways. Rather than presenting just one line, the graph can include a large S-shaped swath (Figure 2.5). Alternatively, it could branch out, beginning at the current time period, indicating a successively widening range of future values for the variable in question.

S-shaped curves also can be used with multiple variables in a way similar to that described for other projection lines. This can quickly become an even more complex issue, and therefore entails greater care in presenting the data, computations, prediction, and associated explanations.

TANDEM PREDICTIONS

A predictive technique that does not rely exclusively on statistical computations is the use of tandem, or associative, predictions. This technique is used when the target variable is difficult to predict, but is strongly associated in a repetitive manner with another variable that is easier to predict. By predicting the easy variable, the presentor can, with some confidence, predict a future value of the target variable.

The association between two variables may be based on statistical experience, common sense, available resources, or theory.

This technique can be illustrated in the prediction of enrollments in various types of educational programs. If a suburb is expected to expand rapidly and attract many new, younger families, a projection line based only on past population may be inadequate and may underestimate school-age populations. However, by projecting school enrollment in tandem with an-

DIAGRAM A   PREDICTED FEASIBILITY FOR LAND DEVELOPMENT

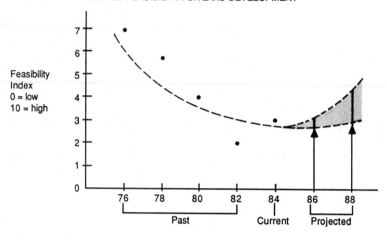

DIAGRAM B

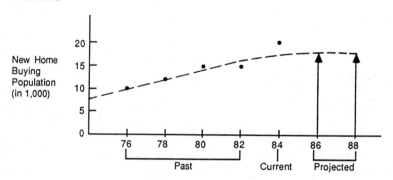

FIGURE 2.4   CREATING AN INDEX   (***Explanation***: an index can be created by combining the values of different variables according to some predetermined formula. Diagrams B to E show data that the presentor has used to create the *feasibility index* shown in Diagram A. It presumes that the audience has accepted the assumptions for creating the index.)

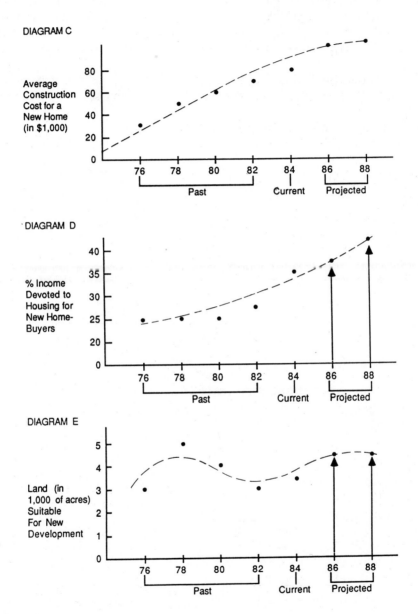

DIAGRAM C

Average Construction Cost for a New Home (in $1,000)

Past   Current   Projected

DIAGRAM D

% Income Devoted to Housing for New Home-Buyers

Past   Current   Projected

DIAGRAM E

Land (in 1,000 of acres) Suitable For New Development

Past   Current   Projected

other variable related to enrollment (such as projected housing construction), a far different, and perhaps more realistic, image may emerge (Figure 2.6). Similarly, predicting the sales of a consumer product based on a tandem prediction for general economic trends rather than past levels of sales may be an effective technique.

Tandem predictions may be appropriate not only because of the strength of the logical association between variables but also because of the time,

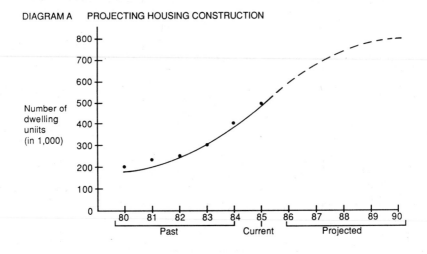

DIAGRAM A   PROJECTING HOUSING CONSTRUCTION

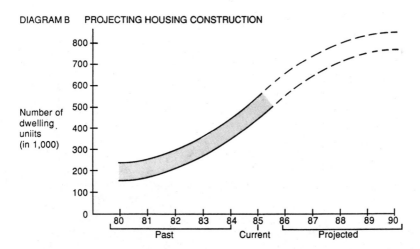

DIAGRAM B   PROJECTING HOUSING CONSTRUCTION

FIGURE 2.5   THE S-CURVE   (**Explanation**: The S-curve is one of a family of shapes that often seem to be intuitively appropriate for predictions (see Figure 2.11). If used, the S-curve must be justified by a strong argument, especially if the complex computations for estimating the type of curve are too time-consuming or do not produce the prediction line with the lowest degree of error. These diagrams show three successively less specific predictions based on the same data.)

money, and expertise available. For example, there may be an existing projection of population changes for a geographic area that was developed by other analysts with a large staff, ample funds, and sophisticated computer models. If population changes for only a portion of that geographic area or for an overlapping area have to be projected and there are few resources available, it may be more reliable to tie the predictions together. In this way, the desired population projection is considered in tandem with the more detailed and thorough projection for the associated area (Figure 2.7).

Presenting tandem predictions creates some difficulties. There must be a clear explanation of why the relationship between the target variable and the associated variable is valid. The audience should be shown a table, chart, diagram, or written explanation justifying the tandem prediction of the two variables, and must be convinced that the relationship between the target variable and the associated variable is likely to continue throughout the time periods in question.

The projections for the associated variable have to be shown and explained to the audience, which may be difficult if the associated variable was projected by other analysts, using data and techniques with which the presentor has no first-hand experience. In such cases, the presentor has to defend some other person's or group's analysis without a full understanding of how it was achieved.

The actual graph or diagram showing the tandem prediction should contain visual evidence of the continuity of the associated variable, the link to

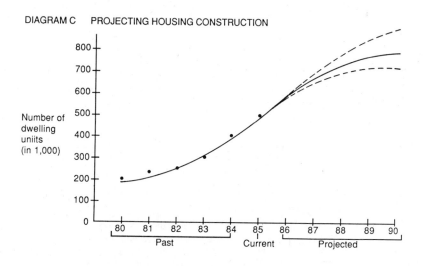

DIAGRAM C   PROJECTING HOUSING CONSTRUCTION

DIAGRAM A   PREDICTED SCHOOL ENROLLMENT

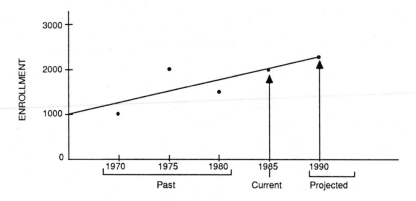

DIAGRAM B   SCHOOL ENROLLMENT PROJECTION
             (BASED ON NEW HOUSING CONSTRUCTION)

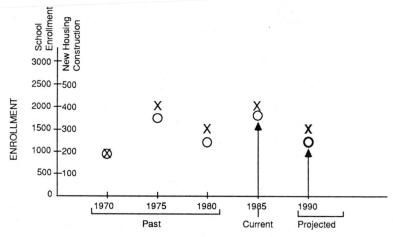

KEY: X = School Enrollment
      O = New Housing Construction

FIGURE 2.6 LINKING TWO VARIABLES IN TANDEM (*Explanation:*   One variable
can be predicted by linking it with another variable that has already been predicted for
some future target date, or which is easier to predict. Diagram A shows a variable
predicted without linkage to an associated variable. Diagram B shows how that
prediction might differ if it were based on an associated, tandem variable.)

the target variable, and the resulting predictions of the target variable. For example, the graph should use a connecting or projection line to show the change over time of the associated variable rather than the target variable. There should be no connecting line between values of the target variable to emphasize to the audience that it is only the associated variable for which continuity over time has been established. However, there should be a series of graphic links (at each time point) indicating the tie between the associated and target variables. Finally, there should be clear denotations of the values of the target variable at different points in time.

For projection line diagrams, it is possible to predict a target variable by using multiple associations. If the target variable is viewed in tandem with two associated variables, the results probably can be shown on one graph or diagram (Figure 2.8). With three or more associated variables, it would be wiser to use several diagrams, with one diagram for each tandem prediction and one summary diagram displaying only the predicted values of the target variable (Figure 2.9).

It also may be possible to show tandem predictions by using wide swaths or branching points. Again, this helps indicate a range of likely future values of the target variable rather than one specific value.

REPEATING THE PAST

The most overlooked predictive technique lies in the assertion that future events will be repetitious of past events. This concept is not very glamorous and often goes against enculturated perceptions that everything is changing. There often is surprise that some phenomena remain the same or that cycles of events previously experienced are repeated.

Showing an audience an analysis that suggests future events will be much like past experience can be unpopular, especially if the audience is focusing on issues where a lack of change implies that a problem will continue. Such audiences are predisposed to accept a prediction for change simply because it holds out the promise of improvement. However, the technique has value where allusion to the past will have a positive effect on the audience or where it is the most accurate and appropriate form of prediction for the problem at hand.

No Significant Change

The simplest form of prediction based on the direct continuation of current experience is that current conditions will persist. That is, no significant increases or decreases are expected in the values of the variables being examined. The key word here is *significant*. For example, there may be a minor change in average income for a community, but no one would expect the precise number to repeat itself down to the last penny.

DIAGRAM A   POPULATION PREDICTION FOR SMALL TOWNS IN REGION A

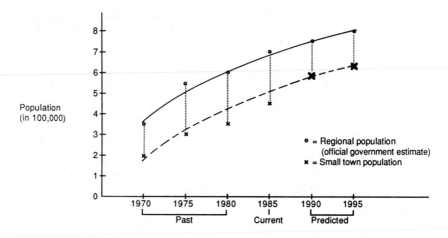

DIAGRAM B   POPULATION PREDICTION FOR SMALL TOWNS IN REGION A

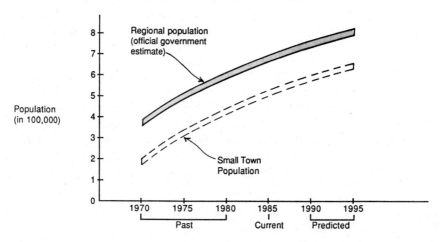

FIGURE 2.7   TANDEM PREDICTIONS FOR MATCHING
VARIABLES   (***Explanation***: Occasionally, it is easier and more reasonable to base a
prediction for one variable on a prediction for a very similar, matched variable. Diagrams
A and B show two versions of a subarea prediction that is based on a population predicted
for a larger base area, that presumably was conducted by using more powerful and
potentially more reliable computational methods.)

DIAGRAM A   COLLEGE ENROLLMENT PROJECTION

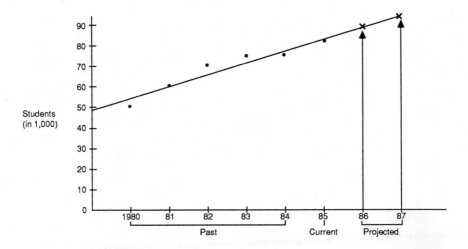

DIAGRAM B   COLLEGE ENROLLMENT PROJECTION

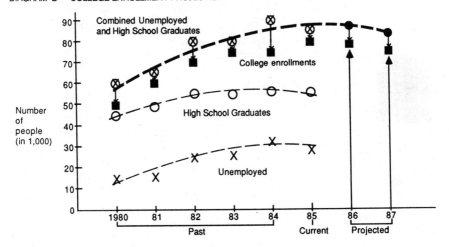

KEY:

■ =TARGET VARIABLE: COLLEGE ENROLLMENTS (IN 1,000)
○ = HIGH SCHOOL GRADUATES (IN 1,000)
X = UNEMPLOYED (IN 1,000)
⊗ = TANDEM VARIABLE: SUM OF UNEMPLOYED AND HIGH SCHOOL GRADUATES
●→ = PROJECTION OF TANDEM VARIABLES
↓ = RELATION OF TANDEM VARIABLE TO TARGET VARIABLE

FIGURE 2.8   TANDEM PREDICTIONS WITH THREE VARIABLES   (***Explanation***: One diagram can accommodate, without visual confusion, roughly three variables. Diagram A shows a projected variable without tandem variables. Diagram B shows a projection with two additional tandem variables. Note that the two diagrams lead to different conclusions.)

DIAGRAM A   PROJECTED COLLEGE ENROLLMENT

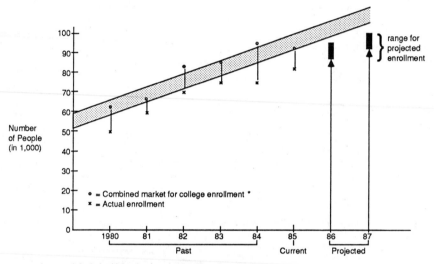

* Includes high school graduates, unemployed and second career/ continuing education population.

DIAGRAM B   HIGH SCHOOL GRADUATES

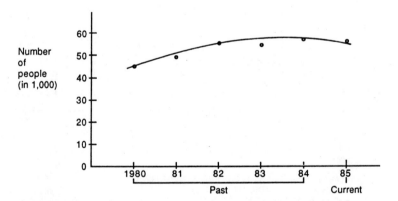

FIGURE 2.9   TANDEM PREDICTIONS WITH MULTIPLE DIAGRAMS   (*Explanation*: When there are too many variables, it is more effective to use one summary illustration (Diagram A) and several supporting illustrations (Diagrams B, C, and D).)

DIAGRAM C   UNEMPLOYMENT

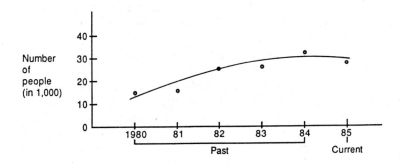

DIAGRAM D   SECOND CAREER AND CONTINUING EDUCATION STUDENTS

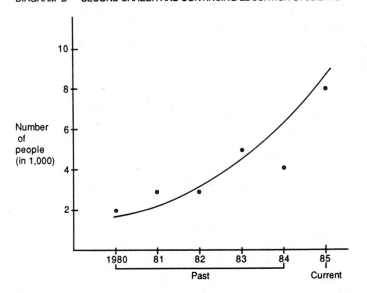

Consequently, the first question the presentor must address is a definition of significant change. This is not a question of statistical significance (defined as the computational formulas whereby an investigator can assert that there is an extremely low probability that two or more values of a variable are truly identical). Rather, it is a question of theory, practice, common sense, and above all the anticipated concerns and attitudes of the intended audience.

For example, in discussing trends in mortality rates, it may be significant if there is a rise of just one point—such as an increase from 6.0 to 7.0 deaths per thousand. After all, that is an increase of almost 15 percent. On the other hand, a change in mortality rates from 6.0 to 6.2 might be considered insignificant if there were past evidence of fluctuations of a similar magnitude or there is reason to believe that changes in the reporting and measuring of data might account for fluctuations in the annual mortality rate of 5 to 10 percent. Therefore, the presentor must establish an argument for that which constitutes significant change.

The values of the variable(s) in question should be shown on the graph or diagram, and the presentor should note the amount of change, and whether it is significant from one time period to the next. A simple way to do this is to draw two lines representing the edges, or boundaries, beyond which a change in the value of the variable would be judged as a significant difference (Figure 2.10).

There may be situations in which one or more of the data points for past time periods do, in fact, indicate a significant difference. That difference has to be noted on the graph and explained to the audience, showing them that significant variations are not part of a time trend, but are instead the result of unique circumstances at that point in time. If, however, such circumstances are not unique or idiosyncratic, the fundamental premise of predicting little or no change should be reexamined.

There are circumstances in which an apparent increasing (or decreasing) trend could form the basis for predicting little or no change. For example, each periodic increase (or decrease) in the trend could be linked to a unique set of events, none of which are likely to occur in the future. This approach could be used in combination with an S-shaped projection line. The presentor could note here that recent rapid increases resulted from a series of unique events that are no longer likely and that imply a leveling off of the value of the variable in question (Figure 2.11).

The Same Rate of Change

There are different ways to interpret the concept of "no significant change." For example, an analyst could argue that there is no significant change in an inflation rate. Some analysts might agree that this implies continuation of the status quo, whereas others would contend that it means a significant increase in the price or cost of an item in question. Using rate of change instead of

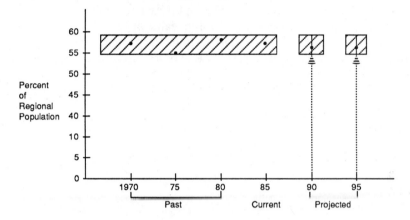

DIAGRAM A   POPULATION OF SMALL TOWNS (AS A PERCENTAGE OF THE
REGIONAL POPULATION)

DIAGRAM B   COLLEGE ENROLLMENT AS A PERCENTAGE OF MARKET POPULATION *

* Includes high school graduates, unemployed and second career/ continuing
education population

FIGURE 2.10   PREDICTING NO SIGNIFICANT CHANGE   (**Explanation**: If a particular issue can be
measured with a variable that appears to remain constant, prediction of "no significant
change" can be convincing. Diagram A, based on the same data as Figure 2.7, is a simple
illustration of this principle. Diagram B, based on the data from Figure 2.9, shows how exceptions
to the status quo can be explained.)

absolute values is a common example of statistical gymnastics. This technique is carried even further when a presenter argues that a particular rate is increasing (or decreasing) at the same pace. In other words, there is constant acceleration (or deceleration) in the value of a variable.

In circumstances where a group of officials need to make a decision based upon predictions of the appreciation of property values, the analyst could develop a projection line based on the absolute value of appreciation, the rate of appreciation, or the periodic change in the rate of appreciation (Figure 2.12). Here, the presenter should pick the statistic that best suits the audience and the

DIAGRAM A   PROJECTED NEW HOUSE CONSTRUCTION

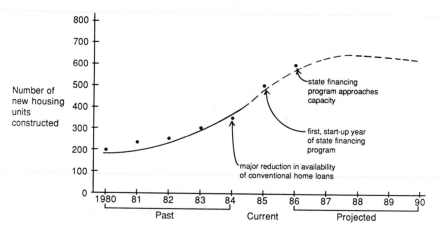

DIAGRAM B   PROJECTED NEW HOUSE CONSTRUCTION

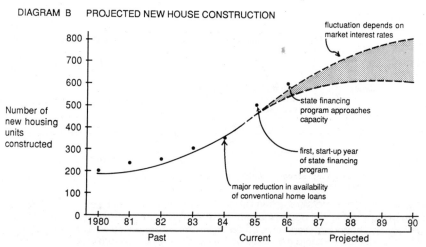

FIGURE 2.11 APPROACHING NO SIGNIFICANT CHANGE (**Explanation**: In some situations, the analyst can argue that a variable that has changed rapidly in the past will now level off, and little or no significant change will occur henceforth. Diagrams A and B show two ways of illustrating the same data.)

Each of these three types of projections could be cast in terms of *no significant change*; an argument to maintain the same level of property investments could be based, in part, on a projection of the rate of appreciation, presuming it too has remained the same. Alternatively, an argument for increasing levels of investment might be better served by projecting the absolute values of appreciation.

Cycles of Change

Predicting events as cyclical is another variant of repeating the past. It is likely to be perceived sympathetically by audiences familiar with clichés about history repeating itself or perhaps about the repetition of themes in fashion design from previous decades. Cycles do not imply that previous occurrences will be repeated in exactly the same manner, but that the significant aspects of the issues under discussion will be repeated. When considering the assertion of a cyclical pattern, the analyst must first judge whether it makes sense from a theoretical and experiential perspective, and not just from a statistical viewpoint.

One of the more common examples of cyclical phenomena is the prediction of annual unemployment levels. Public reports of unemployment rates often say that they are "seasonally adjusted," meaning that there are cyclical increases (or decreases) in certain seasons when reoccurrences of previous seasonal changes are likely. The issue is not the absolute change in unemployment from, say, spring to summer, but rather the change in unemployment that has occurred *in addition* to what the seasonal change would predict.

Cyclical changes are presented in much the same way as other phenomena, with a graph or diagram plotting the points at different time periods. The resulting curve usually is a series of waves, or humps, showing the periodic rise and fall of the variable in question. A common presentation issue is whether or not to present the actual values, or as noted in the example on unemployment, to present *adjusted* values. It may seem more sophisticated and rational to adjust values to account for cyclical change, but it may be inappropriate when the repeated rise and fall of a variable is, in fact, the central issue. In such situations, deemphasizing the fluctuation will mislead the audience. It occasionally is useful to present both actual and adjusted values, possibly on one diagram (Figure 2.13).

There are situations in which statistics alone do not justify a cyclical projection line. For example, in recording the value of a variable, there may be just one wave—one rise and fall—of the numbers. In more extreme situations, just the beginnings of one half a cycle may be visible with no statistical evidence that one full cycle would be completed. However, there may be nonstatistical reasons to presume that a cyclical prediction is valid. Furthermore, as noted previously, the line representing the projection should branch out or become wider after the current time period to imply that the frequency

DIAGRAM A   PROJECTED APPRECIATION IN PROPERTY VALUE

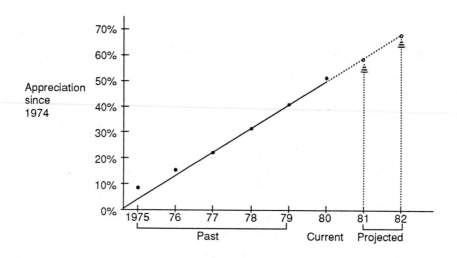

DIAGRAM B   ANNUAL RATE OF APPRECIATION

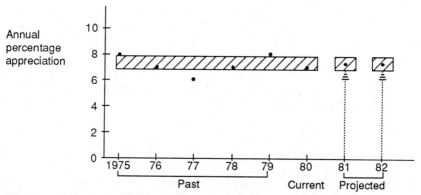

FIGURE 2.12 PREDICTING RATES OF CHANGE (***Explanation***: Dependent on which variable is used to measure a specific issue, the analyst can show an increase (Diagram A), no significant change (Diagram B), or a fluctuating cycle (Diagram C). All of these diagrams are based on an identical data set.)

and depth of the cycle are uncertain (Figure 2.14). In such cases, the analyst should note clearly the specific reasons for the projected upward and downward cyclical movements so that the audience can appreciate the nonstatistical rationale for the prediction.

STATISTICS VERSUS THEORY AND PRACTICE

Controversies over statistical predictions often end with emotional debates over whether the statistics are right, or whether other theoretical or practical approaches are more enculturated in the scientific and technical appeal of statistics (*Numbers don't lie*), which are often referred to as *objective* facts. Alternatively, theories may be labelled *speculative*, whereas practical experience can be described as *arbitrary, anecdotal,* or *subjective*.

These are obviously extremes, and conscientious analysts are usually aware of the subjectivity, qualified validity, and limited reliability of their data. Sound theories and legitimate expertise also are given proper credence in thoughtful analyses, even if they are deemphasized to avoid an "unscientific" appearance.

Attitudes about science, theory, and rational thinking have changed in the last two decades, and subjective analytic judgments are increasingly being given more recognition. Some analysts consider it a necessary evil, while others are beginning to view subjective judgment as a positive contribution.

DIAGRAM C   ANNUAL INCREASE OR DECREASE IN RATE OF APPRECIATION
(IN PERCENTAGE POINTS)

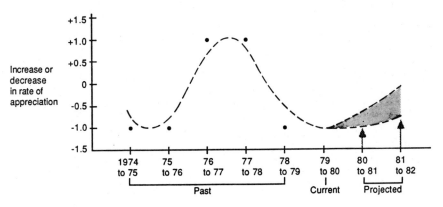

Regardless of this debate, audiences often want to know how theories and expert judgments interface with statistical data in the art of making predictions. One way to present this effectively is to show nonstatistical issues as annotations on projections and graphs. Previous sections of this chapter have illustrated this approach. In every instance, it was the statistics that were the visually dominant element when the statistical projection line was the focus of the audience, and the annotations influencing the direction of that line were secondary. This situation can, of course, be reversed, and may be appropriate when the explanatory statement accompanying the prediction relies heavily on a theory or experiential model that must be accepted by the audience to validate the prediction.

This presentation technique is even simpler than previous examples. The theoretical principles, concepts, or model being used for the prediction are encapsulated and placed directly on the graph or diagram. (A similar process

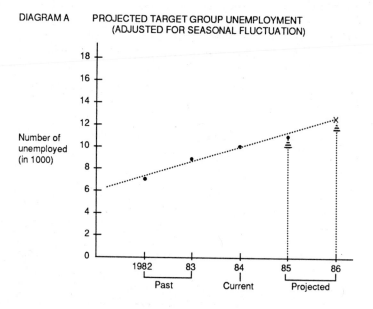

DIAGRAM A   PROJECTED TARGET GROUP UNEMPLOYMENT
(ADJUSTED FOR SEASONAL FLUCTUATION)

Number of unemployed (in 1000)

1982   83   84   85   86
Past   Current   Projected

FIGURE 2.13 REPEATING CYCLES (**Explanation**: Cycles of change, especially for economic issues that have a traditional pattern of seasonal shift, may require special attention. In Diagram A, the variable is shown as "adjusted for seasonal fluctuation." Diagram B, showing both actual and adjusted values, may be more useful if the analyst is concerned with only the "high" values.)

for developing titles is discussed in Chapter Seven.) A time line showing the implications of these statements is drawn, but no statistics are plotted as in the computation of a conventional projection line. They may, however, occur in the explanatory statements and/or illustrate the prediction. It is prudent to draw general ranges of future numerical values rather than overly precise ones (Figure 2.15).

In some cases, it may be more effective to use a series of diagrams. For example, the first diagram might show how a popular theory implies a quick increase in the value of a variable. Another, less well-known model on the second diagram might temper this prediction or, alternatively, imply longer term continuity of the trend. Finally, a summary diagram could show how the prediction implied by the theoretical statements are combined. In other situations, there may be two or more competing theories, each leading to substantially different predictions. Here too separate diagrams for each ap-

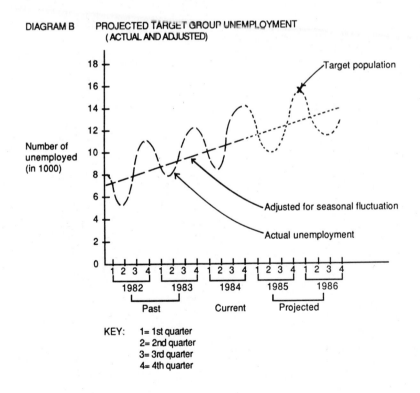

DIAGRAM B   PROJECTED TARGET GROUP UNEMPLOYMENT
             (ACTUAL AND ADJUSTED)

DIAGRAM A   PROJECTED NEW OFFICE CONSTRUCTION

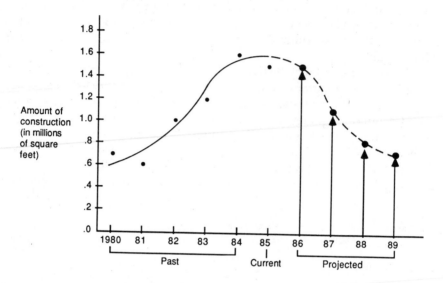

DIAGRAM B   PROJECTED NEW OFFICE CONSTRUCTION

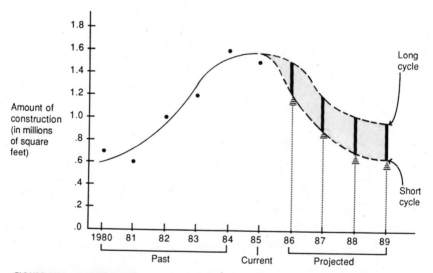

FIGURE 2.14 A SINGLE CYCLE   (***Explanation***: The analyst may want to assert that a particular variable will enter into a "cycle" of change. These diagrams offer two ways to show a projected single cycle.)

DIAGRAM A   PROJECTED NEW CLINIC CONSTRUCTION

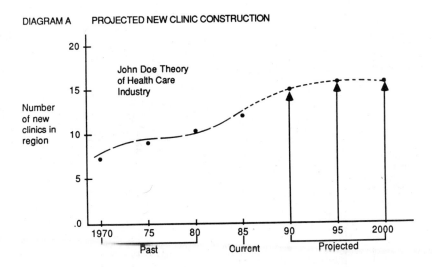

DIAGRAM B   PROJECTED NEW CLINIC CONSTRUCTION

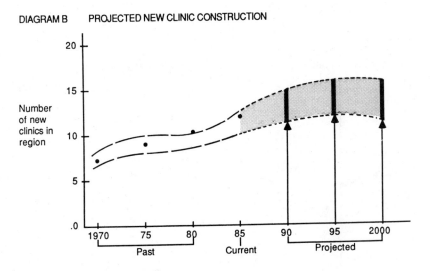

FIGURE 2.15 PREDICTION BASED ON THEORY (***Explanation:*** By using theoretical principles, the analyst can show how a variable will change over time. An overly precise prediction, as is shown in Diagram A, may be harder to justify than a looser or wider ranging prediction, as is illustrated in Diagram B.)

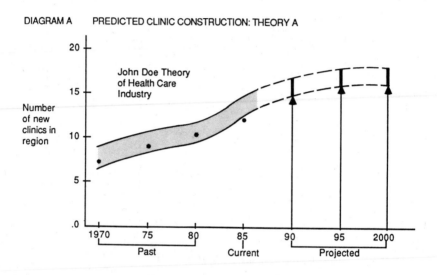

DIAGRAM A   PREDICTED CLINIC CONSTRUCTION: THEORY A

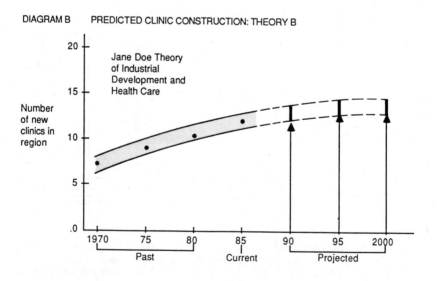

DIAGRAM B   PREDICTED CLINIC CONSTRUCTION: THEORY B

FIGURE 2.16 PROJECTION WITH CONFLICTING THEORIES (*Explanation*: Occasionally, the analyst should present the way in which two conflicting theories each imply different predictions. Diagram A is similar to Figure 2.15. Diagram B shows a competing theory, whereas Diagram C shows a summary of the two preceding diagrams.)

proach and a summary comparative diagram would be effective in explaining the issues to the audience (Figure 2.16).

The last issue concerning predictions based on theory or experience concerns the opportunity for contingent predictions. Projection line techniques follow a mathematical model and cannot be modified to show contingencies based on mathematics alone. With a prediction based on theory or experience, contingencies are more easily presented, and may strengthen an argument, increasing their credibility with the audience.

Contingencies can be portrayed as a series of nodes that branch in two or more directions. Each branch is labelled according to its theoretical or experiential basis (Figure 2.17), and the nodes from which the time lines branch out can be related to specific events or decisions. This technique is especially relevant when it is used jointly with techniques noted in Chapters Four and Five for decision making.

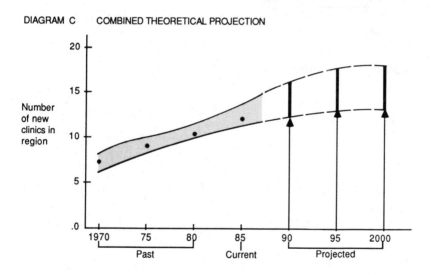

DIAGRAM C   COMBINED THEORETICAL PROJECTION

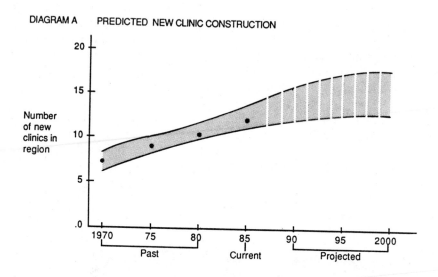

DIAGRAM A   PREDICTED NEW CLINIC CONSTRUCTION

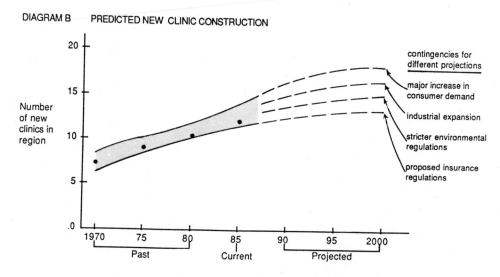

DIAGRAM B   PREDICTED NEW CLINIC CONSTRUCTION

FIGURE 2.17   PREDICTION WITH THEORETICAL CONTINGENCIES   (***Explanation***: Theories can also be combined with expectations regarding future events to create contingent predictions. These two diagrams show the same predictions without contingencies (Diagram A) and with them (Diagram B).)

3

Maps: Spatial Statistics

The practice of presenting data spatially or geographically is relatively undeveloped outside a few disciplines such as geography and urban planning. However, with the advent of computerized graphics and data processing, there have been many advances in presenting geographic data that offer considerable flexibility and facilitate graphic production. However, most of these techniques are intended to increase visual accuracy and clarity, and rarely emphasize the relationship between the presentor and the audience. They tend not to answer questions of what to present or how to aggregate data for effective communication of issues and ideas.

The creation of a map for presenting data is usually treated as a routine task, and the analyst tries to show data objectively and with considerable graphic detail as precisely as possible. This is similar to presenting tables of data with as much detail and as many subcategories as possible. A high degree of detail is appropriate when maps are discussed among analysts who are familiar with the specific situation, but it is less appropriate when dealing with general audiences or other analysts unfamiliar with the situation.

Data should be summarized and organized to emphasize relevant issues and findings and to deemphasize less relevant information. This requires reasonable, subjective judgment. It is less arbitrary, however, than presenting all data in full detail and letting the audience interpret the data as best they can. Letting the audience fend for itself encourages the misinterpretation of data, obfuscation of significant information, and a possible loss of interest. It is far better to summarize data, focus audience attention, and then provide more details in response to inquiries.

GRAPHIC EMPHASIS

Presentations on maps require more graphic expertise and skill than other forms of statistical presentation, as there are more types of graphic decisions to be made. However, the relative visual balance and emphasis can have a more profound impact on the audience's comprehension of the issue. For example, it may be better to communicate overall spatial or geographic patterns than individual statistics for different points or areas.

The presentation of a nonnumeric summary is often necessary, and numeric data can be aggregated into ordinal information, where different subareas are portrayed as having high, medium, or low values. Sometimes, a five- or seven-level scale can be used (Figure 3.1), although specific measurements often should accompany such ordinal gradations.

Qualitative information may be shown concerning how data should be interpreted and/or which subareas or points should be given special attention. Data sometimes are noncomparable, as when the issue being mapped is *major health problems* or *opportunities for economic development*, and the relevant data varies with different subareas (Figure 3.2). (That is, the same variables are not being measured in all cases or areas.) The data therefore are not numerically comparable, even though they all relate to the same qualitative issue. Other, more compatible, data may be effectively presented by using some of the following techniques.

Geographic Orientation

Failing to help observers orient themselves to locations on maps is a common shortcoming. Most maps are oriented so that the compass direction "north" points upwards, but there are some circumstances in which this convention should be modified. For example, a city map might be oriented such that the street grid parallels the vertical and horizontal edges of the map. Thus, the northerly direction may vary as much as 45° from the vertical. There are other situations in which maps are presented sideways to accommodate graphic constraints, such as the standard size of the paper or the board size being used for the presentation. Similarly, there may be local customs that should be followed, where community maps have been presented with the same orientation for many years. Audience comprehension is the key criterion to use in orientation.

In addition to the compass orientation of a map, the audience should be given other cues regarding the location of areas. Maps of urban areas or regions should include landmarks and other well-recognized features, such as principal buildings or monuments, major highways, roads, rivers, and other topographical features. The goal here is to show the audience just enough information so that they "know where they are" when reference is made to different points on the map (Figure 3.3). Even maps that show only overall

MAP A. PERCENT OF NON-TRADITIONAL HOUSEHOLDS BY SUBAREA

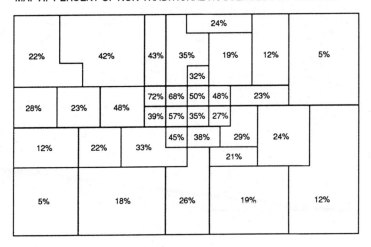

MAP B. PERCENT OF NON-TRADITIONAL HOUSEHOLDS BY SUBAREA

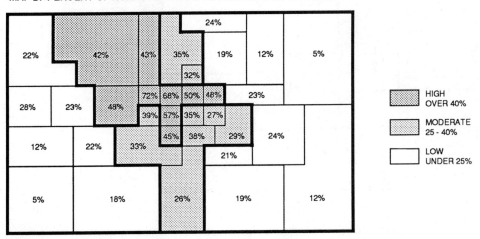

FIGURE 3.1   LEVEL OF MEASUREMENT ON MAPS   (***Explanation***: Different levels of measurement can facilitate or hinder the audience's recognition of spatial patterns. Map A shows data on an interval or numeric scale in which it is difficult to discern a pattern. Map B shows the same data but gives these data secondary emphasis, with primary emphasis focus on the ordinal relationships and on an emerging pattern.)

MAP A. HEALTH CARE PROBLEMS BY SUBAREA

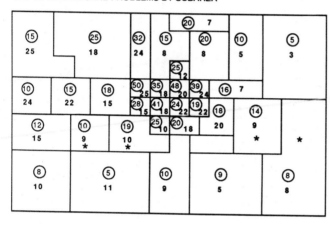

MAP B. AREAS WITH MAJOR HEALTH CARE PROBLEMS

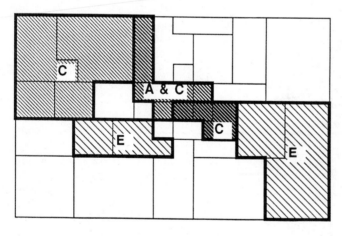

FIGURE 3.2   MAPPING NONCOMPARABLE DATA   (***Explanation***: Raw data for a variety of noncomparable variables can be visually confusing. Map A shows data for three variables. Map B is more
effective because it summarizes the information for the audience. If more data were needed, the
data could, for example, be shown on a separate table along with Map B.)

patterns could contain landmarks and clues for orientation, lack of which might create confusion, misinterpretation, and/or disinterest on the part of the audience.

Gradated Tones

The presenter should consider the application of different tones, line thicknesses, or other markings to differentiate and emphasize points and areas. If five different shadings are used to indicate five different values of a variable, the sequential order of the tones should *match* the sequential order of the values. For example, if the lowest value is shown in white and the highest in black, intermediate values should be shown in gray tones, which get darker as the values increase (Figure 3.4). This seems obvious, but there are numerous errors made, especially in the use of color tones. If yellow and red denote the high and low values of variable A and the color orange is inadvertently selected to represent a different variable B, the audience may be confused because some people will assume orange represents an intermediate value of variable A. Similarly, three red tones may show the first three levels of value, while the last level is shown in purple or brown (perhaps just because the presenter ran out of red tones). An uneven visual pattern then emerges, giving potentially unwarranted emphasis to the last category. These anomalies also can occur when only black, white, and gray tones are used (Figure 3.4).

The visual intensity of gradated tones of the map should also be checked to ensure that it reflects the major issues. It is easy to make mistakes. For instance, data may be aggregated for a map divided into five subareas, one of which comprises three-fourths of the area. This occurs when showing a dense urban area surrounded by a large suburban ring that is treated as one subarea. Whichever tone is selected, the large subarea will dominate the map. If the large subarea is of lesser relevance or just equal relevance to the other areas, it should not be given a dominant color or tone. Consequently, tones and gradations cannot be selected without considering the size of the subareas to be toned. Effective communication to the audience requires simultaneous consideration of tone sequences and the size and location of areas to which they are applied.

Line Weights

One method of emphasizing gradated values is to vary the types of lines and markings used to draw boundaries or special points. For example, subareas of greater importance can be drawn with a heavier boundary, while focal points can be shown with a darker point or circle. Different types of lines could be used to denote two or three types of areas or points; a heavy solid line could be used as the boundary for the most important areas (or points), a heavy dashed line for secondary areas (or points), and a light solid line for all other

MAP A. SITE OPTIONS AND RELATED FEATURES FOR NEW DEVELOPMENT

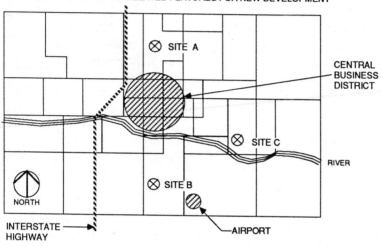

MAP B. PROPOSED ARTS DISTRICT AND RELATED FACILITIES

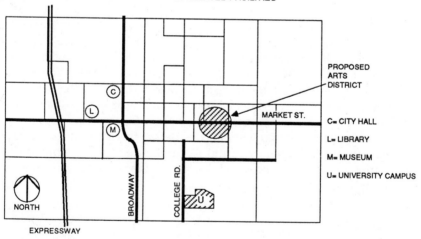

FIGURE 3.3 ORIENTATION ON MAPS (**Explanation**: Maps should contain key landmarks and features to help the audience relate the information being presented to their "mental map" of the area being studied. Map A, for example, might be used in a presentation involving analysis of three sites for a new commercial development. Map B might be used in an analysis regarding cultural facilities.)

MAP A. TARGET AREAS FOR ECONOMIC DEVELOPMENT PROGRAM

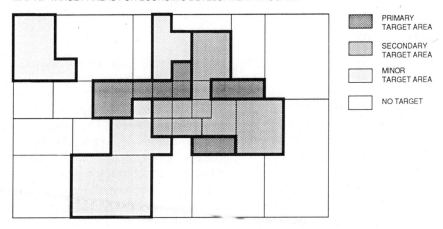

PRIMARY
TARGET AREA

SECONDARY
TARGET AREA

MINOR
TARGET AREA

NO TARGET

MAP B. TARGET AREAS FOR ECONOMIC DEVELOPMENT PROGRAM

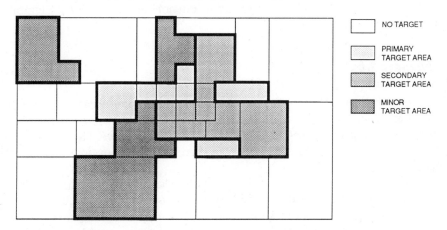

NO TARGET

PRIMARY
TARGET AREA

SECONDARY
TARGET AREA

MINOR
TARGET AREA

FIGURE 3.4 MAPS AND TONES (***Explanation***: The sequence of tones on a map is critical to audience comprehension. Maps A and B show identical information. Map A is graphically superior because it treats the category of *no target area* as the logical last step in a descending order, and the *primary target area* as the first step. Map B confuses the pattern, making it difficult to see the descending order of importance.)

(Figure 3.5). The presentor may also decide to reserve the use of different lines or tones for a few special points or areas, leaving most of the map as a neutral background.

Linked Maps

Sometimes, the concentration of data points can vary greatly from one part of the map to another. For example, a display of data on a state or regional map may have one-half of the relevant data concentrated in densely populated areas that comprise only one-tenth of the surface area (Figure 3.6). One technique that effectively addresses this is the use of blowups or enlargements of portions of the map where heavier concentrations of data are displayed.

The principle behind this technique can be extended to other situations, as where data has been aggregated for an entire city, but there are really only three subareas of interest. It may be inappropriate to show all the data on one large city map, so an alternative may be to indicate the three subareas on one small city map, and then use adjacent enlargements of the subareas to display the information (Figure 3.7).

In addition to enlargements, other visually adjacent notes, diagrams, and statistical data can be keyed into a map by using arrows and lines. The visual order of adjacent notes, connecting lines, and arrows should be simple and direct, facilitating audience understanding, but not interfering with the legibility of the map (Figure 3.8).

DISCRETE AREAS

Conventional presentations usually subdivide geographic areas into discrete subareas, where boundaries do not overlap and no areas are excluded. The subareas are mutually exclusive and collectively exhaustive. This is a simple situation to depict graphically; however, there are several options and issues open to the presentor.

There is always a question as to where to draw the boundaries. Data usually have been aggregated according to subareas such as census tracts, zip codes, or local governmental units. These often are political boundaries that may not be directly relevant to the issues under discussion. A common difficulty occurs when the analyst wants to show data for subareas with one set of boundaries, but the data have been collected in relation to other boundaries. For example, if data aggregated by census tracts or zip codes do not coincide with a set of intended neighborhood boundaries, the analyst may use the census tract or zip code boundaries as *approximations* of neighborhood boundaries. This seems objective but in fact misrepresents the true but unknown values of the variable for each neighborhood.

Approximations should only be used when the presentor judges this to be a reasonable approach, and one option here is to use two sets of boundaries.

One set represents the boundaries of available data, whereas the second represents the relevant subareas. The audience can then see the difference and judge for themselves the value of the approximation (Figure 3.9). The delineation of both types of areas should be clearly visible, but the target subareas should be visually dominant. A similar approach omits the boundaries used for data collection, shows only the boundaries of the relevant subareas, and clearly notes that the statistics are approximations.

MAP A. TARGET AREAS FOR ECONOMIC DEVELOPMENT PROGRAM

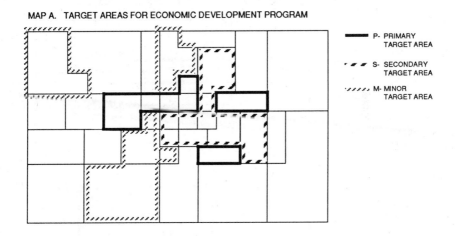

P- PRIMARY
TARGET AREA

S- SECONDARY
TARGET AREA

M- MINOR
TARGET AREA

MAP B. AREAS WITH MAJOR HEALTH CARE PROBLEMS

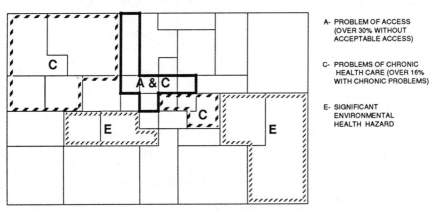

A- PROBLEM OF ACCESS
(OVER 30% WITHOUT
ACCEPTABLE ACCESS)

C- PROBLEMS OF CHRONIC
HEALTH CARE (OVER 16%
WITH CHRONIC PROBLEMS)

E- SIGNIFICANT
ENVIRONMENTAL
HEALTH HAZARD

FIGURE 3.5 MAPS AND LINE WEIGHTS (**Explanation**: Sometimes it is preferable to differentiate areas by using lineweights instead of tones. For example, the tones may obscure other written information or may create an imbalance in the intended visual emphasis. Map A uses lineweights to show the same data as in Figure 3.4. Map B shows the same data as in Figure 3.2.

Another strategy involves modification of the statistics by interpolating the numerical values to fit the desired boundaries. For example, if a census tract is split such that one-third is in neighborhood *A* and two-thirds is in neighborhood *B*, one-third of the quantity is attributed to *A* and two-thirds to *B*. This assumes that the variable in question is evenly distributed throughout the census tract. The wisdom of such assumptions depends on the variable and

TARGET AREAS FOR ECONOMIC DEVELOPMENT

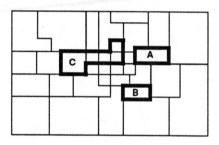

NUMBERS INDICATE PERCENT OF UNEMPLOYED SKILLED LABOR IN RESIDENTIAL BLOCKS

NON-RESIDENTIAL AREA

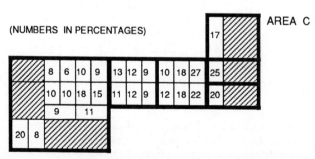

FIGURE 3.6 SUBAREA ENLARGEMENTS **(Explanation:** If a few subareas are of special interest, they can be effectively presented by reducing the size of the base map and then using enlargements of the target areas. This illustration is based on Figure 3.4.)

area in question, and creates some obvious problems in defending the validity of the statistics. On the other hand, it may increase their relevance; in especially critical situations, the analyst can present both the original and the interpolated data as two ways to approximate the true value (Figure 3.10).

The interpolation of data leads to a larger issue concerning presentation of potential errors of measurement. The audience should be shown some

TARGET AREAS FOR ECONOMIC DEVELOPMENT

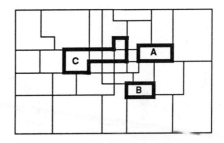

<u>KEY</u>

UNEMPLOYED SKILLED LABOR
IN RESIDENTIAL AREAS

● 10% OR MORE

⊘ UNDER 10%

E.I.- EXISTING INDUSTRY
N.I.- NEW INDUSTRY
(LAST 3 YEARS)'
L- AVAILABLE LAND

C - COMMERCIAL NODES

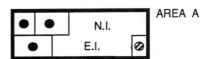

AREA A

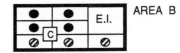

AREA B

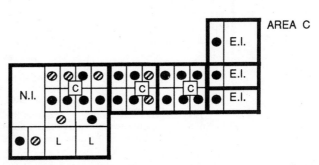

AREA C

FIGURE 3.7   ENLARGEMENTS WITH EXPANDED DATA (*Explanation*: Enlargements of subareas can be used to show a wide variety of issues. This illustration expands Figure 3.6 to add additional information that may be relevant to the audience.)

TARGET AREAS FOR ECONOMIC DEVELOPMENT

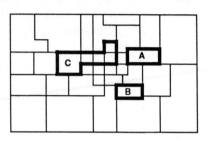

KEY

UNEMPLOYED SKILLED LABOR
IN RESIDENTIAL AREAS

●   10% OR MORE

⊘   UNDER 10%

E.I.- EXISTING INDUSTRY
N.I.- NEW INDUSTRY
     (LAST 3 YEARS)'
L-   AVAILABLE LAND

[C] - COMMERCIAL NODES

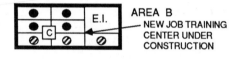

AREA A

60 NEW JOBS
EXPECTED OVER
NEXT 5 YEARS

AREA B

NEW JOB TRAINING
CENTER UNDER
CONSTRUCTION

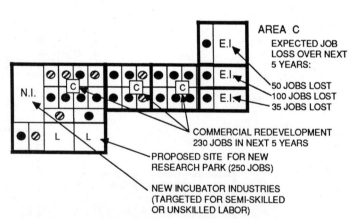

AREA C

EXPECTED JOB
LOSS OVER NEXT
5 YEARS:

50 JOBS LOST
100 JOBS LOST
35 JOBS LOST

COMMERCIAL REDEVELOPMENT
230 JOBS IN NEXT 5 YEARS

PROPOSED SITE FOR NEW
RESEARCH PARK (250 JOBS)

NEW INCUBATOR INDUSTRIES
(TARGETED FOR SEMI-SKILLED
OR UNSKILLED LABOR)

FIGURE 3.8   ANNOTATED ENLARGEMENTS (***Explanation***: Enlargements of
subareas can also be made more effective with explanatory notes, annotations, and
related information. This illustration builds on the data presented in Figures 3.6 and 3.7.)

indication of the validity and reliability of the data. Statistical estimates for some areas may be very precise, but may be based on a variety of assumptions that are debatable, as measurements are, by definition, taken prior to the time of presentation. This is a problem for most statistical presentations, but occurs most frequently on maps based on census data, which often are used and presented as much as five to ten years after the original observations have been made. Furthermore, data may be based on a sample of an area's population rather than on a full or 100 percent enumeration. There is, therefore, a range of error, or *confidence interval*, within which lies the true value of the statistic.

Variations in estimates occur when the availability of data varies from one subarea to the next; data for one state or region may be current, whereas for another the data are two years old. Similarly, two different operational definitions may have been used to gather data in two different cities. For these reasons, the margin for error may differ from one subarea to another, and it may be wiser to present these inherent ambiguities rather than to obscure them by presenting only one set of estimates as being the most accurate. For example, the statistics for each area could be shown as a range with high and low estimates (Figure 3.11), and the presentor can explain that estimates for some areas are more accurate than others. If a particularly important subarea has a wide margin of error, the audience should be made aware of it. Similarly, if added steps were taken to reduce the margin of error for key subareas, this can heighten the audience's appreciation for the work and increase their confidence in the more important estimates.

OVERLAPPING AREAS AND EDGE CONDITIONS

There are circumstances in which it may be inappropriate to base presentations on discrete, nonoverlapping areas. Presentations frequently entail overlapping catchments, or service and/or market areas. Even if discrete boundaries are possible, it may be unwise to draw them, especially if the location of the boundary is a critical decision involving the audience. In such instances, it may be better to show overlapping, ambiguous, or indecisive boundaries.

Showing overlapping areas is analogous to showing overlapping/intersecting sets (see the section entitled "Sets" in Chapter One). For example, boundaries may be drawn as simple geometric forms (such as circles, ovals, or rectangles) that communicate the general shape of the relevant subarea and, at the same time, imply the problematic nature of showing fixed boundaries. The statistical data contained within the boundary can show a range of possible values corresponding to different assumptions and definitions.

The size or degree of overlap between areas depends on the issues being discussed. If the audience is likely to be concerned that some portions or subareas have been omitted, the boundaries should be larger and more inclusive. This approach avoids the error of excluding areas that the audience may want to be included (Figure 3.12).

PERCENT POPULATION CHANGE (1975 TO 1985) FOR
NEIGHBORHOOD AREAS "A" THROUGH "L"

MAP A. POPULATION CHANGE AGGREGATED USING CENSUS TRACT BOUNDARIES

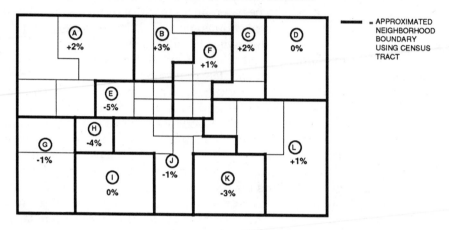

MAP B. ASSUMED POPULATION CHANGE BY NEIGHBORHOOD BOUNDARIES

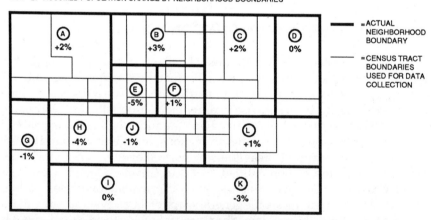

FIGURE 3.9 DATA BOUNDARY PROBLEMS (**Explanation**: Available data may not be organized using the same boundaries with which the analyst is concerned. If the problem is sufficiently important the analyst might present two maps, as is shown here, so that the audience can see the correspondence (or lack of it) between the *data collection* boundaries (Map A) and the *intended* boundaries (Map B).)

MAP A: PERCENT POPULATION CHANGE BY CENSUS TRACTS AND BY
NEIGHBORHOOD BOUNDARIES ("A" THROUGH "L")

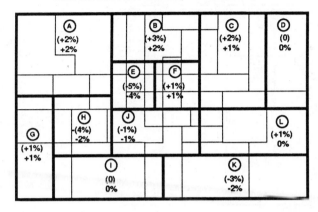

MAP B: PERCENT POPULATION CHANGE AGGREGATED BY CENSUS TRACT
BOUNDARIES FOR NEIGHBORHOODS "A" THROUGH "L"

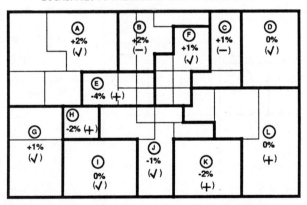

━━━ = APPROXIMATE NEIGHBORHOOD BOUNDARY USING CENSUS TRATS
00% = PERCENT POPULATION CHANGE USING CENSUS TRACT BOUNDARIES

(+) = ESTIMATED PERCENT FOR ACTUAL NEIGHBORHOOD BOUNDARIES IS HIGHER
(−) = ESTIMATED PERCENT FOR ACTUAL NEIGHBORHOOD BOUNDARIES IS LOWER
(√) = ESTIMATED PERCENT FOR ACTUAL NEIGHBORHOOD BOUNDARIES IS APPROXIMATELY SAME

FIGURE 3.10 PRESENTING INTERPOLATIONS ON MAPS (*Explanation*: When there are
sufficient differences between data aggregated according to available sources versus
interpolation of these data, the analyst may wish to present both sets of figures. Maps A and
B, both based on Figure 3.9, offer two options for presenting actual and interpolated data.)

MAP A: PERCENT POPULATION CHANGE (1975 TO 1985) FOR
NEIGHBORHOOD AREAS ("A" THROUGH "L")

			KEY:

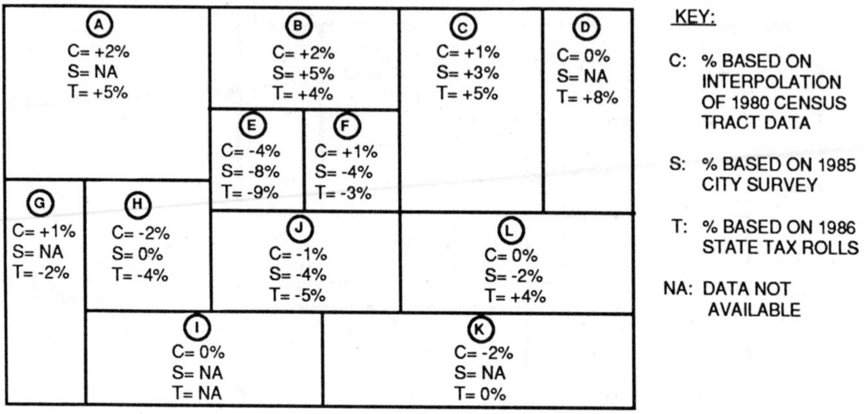

MAP B: PERCENT POPULATION CHANGE (1975 TO 1985) FOR
NEIGHBORHOOD AREAS ("A" THROUGH "L")

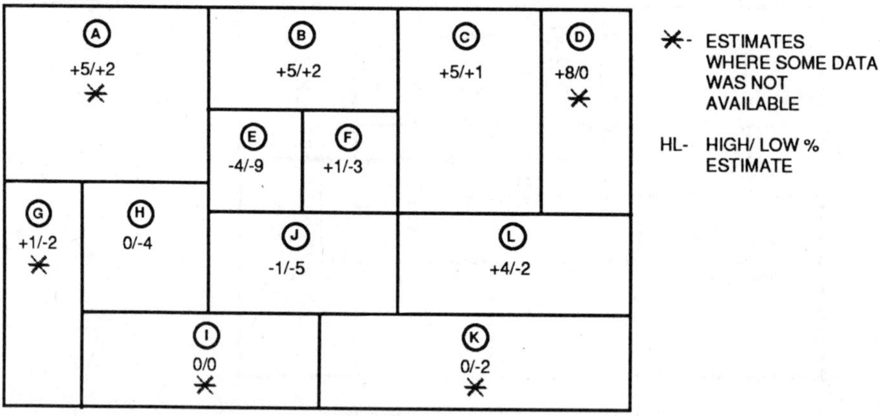

NOTE: ESTIMATES ARE BASED ON INTERPOLATED
DATA FROM 1980 CENSUS, 1985 CITY SURVEY
AND 1986 STATE TAX ROLL INFORMATION.

FIGURE 3.11 MAPS WITH RANGES *(Explanation:* When several different data sources are used, each with different limitations, the analyst may wish to present the data as less exact, in the form of numeric intervals or ranges. Map A illustrates a more precise and complex option whereas Map B portrays a simpler presentation of the same data that is less precise. Both illustrations are related to Figure 3.9.)

Alternatively, the size of subareas may be narrowed so that there are gaps between them, which look like thick lines between areas. They, too, imply ambiguous and indecisive boundaries. This approach avoids the error made by overlapping boundaries, ensuring that no areas are included within a boundary that should be excluded (Figure 3.13).

Obviously, both types of errors — being overly exclusive or inclusive — should be limited, but statistics are not usually sufficiently valid or reliable to avoid both potential pitfalls. Consequently, an effective presentation may entail both overlapping and separated subareas. This creates obvious difficulties in aggregating and interpolating data, but can, nevertheless, create a far more relevant and meaningful presentation. It tells the audience how the data relate to the issues and helps them comprehend the assumptions, limits, and inherent strengths and weaknesses in the analysis.

Graphic presentation of overlapped and separated subareas can be complex, however. The use of tones and shades may be ineffective if too much emphasis is placed on the intersection of overlapped areas. One solution uses stripes instead of tones, where intersections of overlapped areas have alternating stripes that correspond to the tone or color of the two respective overlapping areas (Figure 3.14). This technique must be used cautiously, or it may produce a visual circus rather than a clarification of the situation.

Frequently, when areas overlap, only the boundaries of the areas are shown. Different line weights or line colors can be used so that areas of overlap are easily discerned. Points of intersection sometimes are given special emphasis, or the lines that define the intersections are shown as dashed rather than solid (Figure 3.15). When subareas are separated rather than overlapped, the gap between them may be given special tone, striping, or similar denotation. The presentor should experiment with different techniques in showing overlap and should use the most effective one for the statistics in question.

FOCAL POINTS

Some presentations focus on points or targets rather than subareas. There may be a set of specific points at which special conditions occur, services are provided, or activities take place. For example, in an urban area data may be linked directly to specific point locations such as hospitals, social service centers, libraries, cultural facilities, or schools. The data might be the amount of services provided, costs, revenues, or similar issue. The analyst may wish to present these statistics in association with particular points on the map rather than larger subareas. This usually is shown by including the data in a circle or rectangle centered on the relevant point locations. If there is an excessive concentration of points so that the data cannot be shown on the map, a code may be used with a separate table. Alternatively, a series of lines or arrows can be drawn and the information can be displayed adjacent to the map. Different symbols occasionally can be used to denote different types of points, such as public versus private institutions, new versus old facilities, or high-volume versus low-volume activities (Figure 3.16).

POINTS AND PATTERNS

Presentation difficulties may arise when data aggregated for one point also imply areal (or nonpoint) patterns around that focal point. For example, the

MAP A: TARGET AREAS FOR ECONOMIC DEVELOPMENT PROGRAM

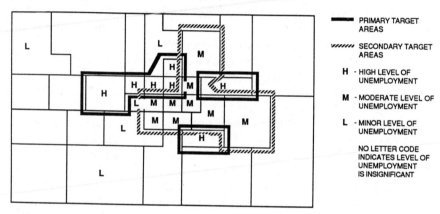

MAP B: TARGET AREAS FOR ECONOMIC DEVELOPMENT PROGRAM

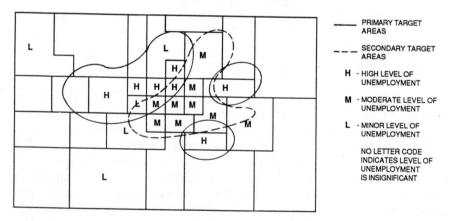

FIGURE 3.12 FUZZY BOUNDARIES—OVERLAPS (*Explanation*: Occasionally data are used as criteria to select or define subareas on maps. To avoid the error of excluding marginal subareas the analyst should draw overlapping boundaries. Two options are shown here—one more precise and complex (Map A) and one simpler and less precise (Map B). Both are related to Figure 3.4.)

number of patrons for a retail center, the number of persons using a park, or the number of clients at a health center all may be expected to cluster around their respective point locations in some spatial pattern. Patterns of concentration and dispersal may vary considerably and take on different shapes and intensities, and portraying these patterns often is a significant issue deserving special attention in a presentation.

Computing and presenting concentrations and/or dispersals of data around a central point can be deceptive, as computations usually involve

MAP A: TARGET AREAS FOR ECONOMIC DEVELOPMENT PROGRAM

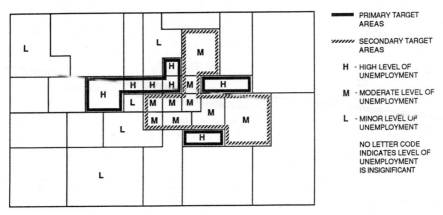

MAP B: TARGET AREAS FOR ECONOMIC DEVELOPMENT PROGRAM

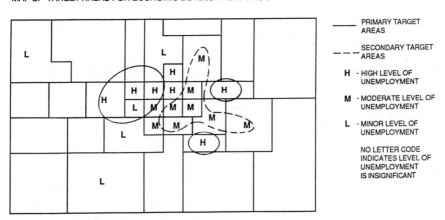

FIGURE 3.13 FUZZY BOUNDARIES—GAPS (**Explanation:** When the presentor wants to avoid the error of including marginal areas, boundaries should be narrowed so that gaps occur. This example is the reverse of the situation in Figure 3.12. Map A is the more precise example, and Map B is the less precise but visually clearer example.)

MAP A. MAJOR HEALTH CARE PROBLEMS BY SUBAREA

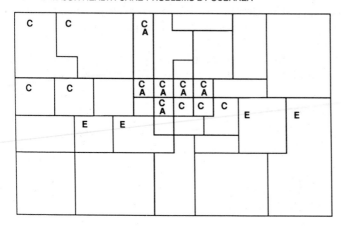

C - CHRONIC HEALTH
CARE (OVER 16%
WITH CHRONIC
PROBLEMS)

A - PROBLEMS OF
ACCESS (OVER
30% WITHOUT
ACCEPTABLE
ACCESS)

E - SIGNIFICANT
ENVIRONMENTAL
HEALTH HAZARD

MAP B. MAJOR HEALTH CARE PROBLEMS BY SUBAREA

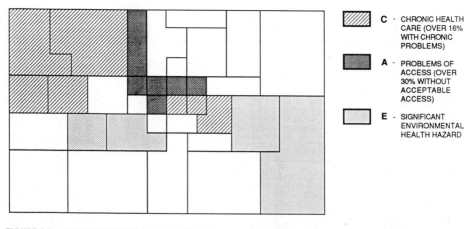

C - CHRONIC HEALTH
CARE (OVER 16%
WITH CHRONIC
PROBLEMS)

A - PROBLEMS OF
ACCESS (OVER
30% WITHOUT
ACCEPTABLE
ACCESS)

E - SIGNIFICANT
ENVIRONMENTAL
HEALTH HAZARD

FIGURE 3.14 TONES AND STRIPES FOR OVERLAPS (***Explanation:*** In some instances the use of tones and stripes is an effective way to show overlapping boundaries, especially for noncomparable issues. Map A is based on the data of Figure 3.2, but the criteria for emphasizing subareas are slightly different leading, therefore, to a different presentation technique, as is shown in Map B.)

straightforward percentages. The *percent of users per area* who use a facility at point *X* may be computed for three areas, *A*, *B*, and *C*, where the percentages are 10, 20, and 25 percent, respectively. On the other hand, the *percent of users per facility* who come from areas *A*, *B*, and *C* may be 50, 35, and 15 percent — in other words, the way in which percentages are defined and measured makes a dramatic difference (Figure 3.17). There are several texts in geography, urban planning, economics, and related disciplines that provide different theoretical arguments, indices, ratios, and measurements of such phenomena, but the issue here is how to present geographic and spatial patterns effectively.

There usually is no single comprehensive technique for showing all geographic patterns on one map, and the analyst has to select the most relevant way of presenting a limited view of the data. Where there are several areas to show, each of which contain people who use different facilities, the analyst may wish to illustrate the geographic pattern of facility utilization. It is possible to decide that each area is a *principle user* for one facility, and arrows, lines, or shading techniques can visually tie each subarea to the one facility with which it is matched. The operational definition of *principle user* is critical here. Another option would be to draw a circle, ellipse, or polygon around each point where the edge of the shape denotes the approximate *sphere of influence, market area,* or *catchment* for a facility. Operational definitions are again critical (Figure 3.18).

A related option, which may be effective with only a few focal points, would be to draw a series of concentric rings or polygons about each focal point. The ring closest to each point would be drawn in the heaviest line, and the line weights would become progressively lighter. The first ring could denote a primary area, the next ring a secondary area, and so on (Figure 3.19). The rings should be simple shapes with easily understood geometries and, as some precision may be lost, the rings should be labelled clearly as approximations.

Complex patterns occasionally can be shown without direct links to the focal points. For example, each subarea could be coded and toned accordingly as to whether it contains a heavy concentration of users at only one, two, or three facilities. Focal points would be shown, but not directly linked to areas. These linkages would have to be explained elsewhere. Nevertheless, a relatively complex geographic pattern can be simply portrayed in this way (Figure 3.20).

It is essential for the presentor to define operational terms and create appropriate titles for the concepts. Titles such as *pattern of concentration, primary concentration, degree of dispersion,* and *principal users* can create different images for audience members, and computation examplars may be necessary.

The analyst should also experiment with the computational sensitivity of operational terms. For instance, changing the definition of primary market area from 25 to 30 percent of a subarea may create a dramatic change in the visual pattern (Figure 3.21). When situations like this arise, the presentor should consider modifying operational definitions to reflect a more consistent pattern that is less sensitive to minor computational shifts. When the patterns are too complex and/or there are too many focal points, multiple maps, as described in the next section, should be considered.

MAP A. MAJOR HEALTH CARE PROBLEMS BY SUBAREA

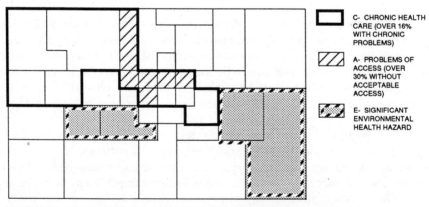

FIGURE 3.15   LINE WEIGHTS FOR OVERLAPS AND INTERSECTIONS   (**Explantion**: Lineweights can be used instead of tones or in conjunction with tones to portray the issues. Map A presents the same data and issues as Figure 3.14 using lineweights. Map B also presents the same data using lineweights and tones in a different fashion.

PATIENT VISITS TO MAJOR HEALTH CARE CLINICS
IN CITY NEIGHBORHOODS

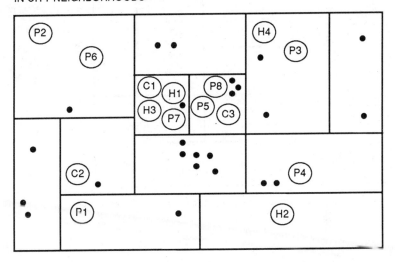

KEY: (XX) = LOCATION OF HIGH VOLUME CLINICS WITH OVER 8,000 VISITS
             PER YEAR (C= CITY, H= HOSPITAL CLINIC, P= PRIVATE CLINIC)

        ● = OTHER CLINICS WITH LESS THAN 8,000 VISITS PER YEAR

	C- CITY CLINICS	H- HOSPITAL AFFILIATED CLINICS	P- PRIVATE FREE-STANDING CLINICS
NUMBER OF VISITS (IN 1000's)	C1 = 45	H1 = 120	P1 = 115
	C2 = 20	H2 = 85	P2 = 92
	C3 = 12	H3 = 50	P3 = 72
		H4 = 50	P4 = 61
			P5 = 33
			P6 = 20
			P7 = 19
			P8 = 15

FIGURE 3.16 FOCAL POINTS *(Explanation*: Geographic data are often associated with a point on a map rather than a subarea. This illustration shows one option of combining a map with a related table. If the map were larger it would be feasible to include all of the data on the map and to eliminate the need for the table.)

MAP A: PERCENT OF NEIGHBORHOOD POPULATION USING HIGH VOLUME CLINICS (OVER 70,000 VISITS/ YR.)

(P2)	H1-22 H2-18 P1-12 P2-40 P3-8	H1-28 H2-8	P1-9 P2-35 P3-20	H1-22 H2-20 (P3)	H1-20 H2-22 P1-7 P2-9 P3-42	

H= HOSPITAL CLINIC

P= PRIVATE CLINIC

⬤ = SITE LOCATION FOR HOSPITAL (H) CLINICS AND PRIVATE (P) CLINICS

Map A regions:
- (P2)
- H1-22 H2-18 P1-12 P2-40 P3-8
- H1-28 H2-8 / P1-9 P2-35 P3-20
- (H1) H1-40 H2-7 P1-9 P2-30 P3-14 / H1-35 H2-7 P1-10 P2-26 P3-22
- H1-22 H2-20 (P3) / H1-20 H2-22 P1-7 P2-9 P3-42
- P1-5 P2-8 P3-40
- H1-17 H2-15 P1-35 P2-23 P3-10
- H1-22 H2-13 P1-42 P2-13 P3-10
- H1-36 H2-30 P1-21 P2-5 P3-8
- H1-15 H2-30 P1-20 P2-15 P3-20
- (P1) H1-12 H2-28 / P1-52 P2-6 P2-2
- H1-20 H2-35 (H2) P1-14 P2-12 P2-19

SAMPLE EXPLANATION: 17% OF PERSONS IN THIS AREA USE H1,
15% " " " " " " H2,
35% " " " " " " P1,
23% " " " " " " P2,
10% " " " " " " P3,
TOTAL 100%

MAP B: PERCENT OF CLINIC VISITS FROM EACH NEIGHBORHOOD

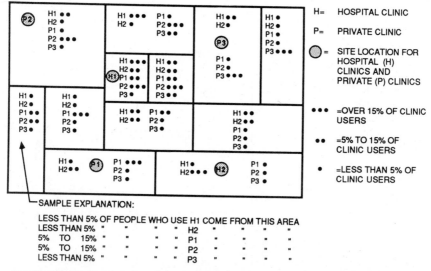

H= HOSPITAL CLINIC

P= PRIVATE CLINIC

⬤ = SITE LOCATION FOR HOSPITAL (H) CLINICS AND PRIVATE (P) CLINICS

●●● =OVER 15% OF CLINIC USERS

●● =5% TO 15% OF CLINIC USERS

● =LESS THAN 5% OF CLINIC USERS

SAMPLE EXPLANATION:

LESS THAN 5% OF PEOPLE WHO USE H1 COME FROM THIS AREA
LESS THAN 5% " " " " H2 " " " "
5% TO 15% " " " " P1 " " " "
5% TO 15% " " " " P2 " " " "
LESS THAN 5% " " " " P3 " " " "

FIGURE 3.17 POINTS AND PERCENTS (*Explanation*: Frequently the analyst wishes to associate percentages with point locations. It can be confusing to the audience unless the analyst illustrates what the percentages actually measure. Map A and Map B offer two examples of relating percentages to points. At first glance, the issues may seem identical, but the sample explanations illustrate the difference. See Figure 3.18 for more effective ways to summarize these data.)

MULTIPLE MAPS AND VARIABLES

Multiple maps most frequently are used when the information is too complex to be shown solely on one map, or when multiple variables need to be displayed. Multiple issues sometimes can be shown on one map if there are only a few issues, or if separate issues relate to separate parts of the map. Even

MAP A: PRIMARY MARKET AREAS * FOR HIGH
VOLUME CLINICS

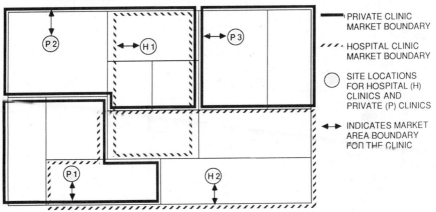

MAP B: PRIMARY MARKET AREAS* FOR HIGH
VOLUME CLINICS

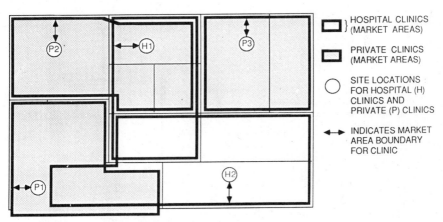

* MARKET AREAS INCLUDE ONLY THOSE NEIGBORHOODS (a) IN WHICH
25 % OR MORE OF EACH NEIGHBORHOOD POPULATION USES THE
ASSOCIATED CLINIC AND (b) WHERE THE NEIGHBORHOODS COLLECTIVELY
ACCOUNT FOR OVER 50% OF THE VISITS TO THAT CLINIC

FIGURE 3.18 POINTS AND CATCHMENTS (*Explanation*: Focal points frequently are associated with catchment or market areas. Both of these maps interpret the data of Figure 3.17 to portray the same catchments in different ways.)

in these special cases, multiple maps may be preferable to ensure clarity of communication.

A conventional approach involves using one map to summarize key findings along with a series of supplementary maps. The summary map may show data selectively in different areas (Figure 3.22), and may have focal

PRIMARY AND SECONARY MARKET AREAS*
FOR HOSPITAL CLINICS

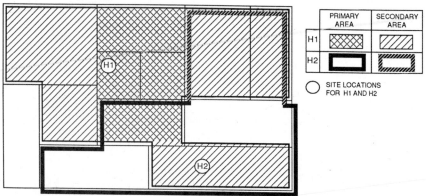

* PRIMARY MARKET AREAS INCLUDE ONLY THOSE NEIGHBORHOODS (a) IN WHICH
25% OR MORE OF EACH NEIGHBORHOOD POPULATION USES THE ASSOCIATED
CLINIC AND (b) WHERE THE NEIGHBORHOODS COLLECTIVELY ACCOUNT FOR OVER
50% OF THE VISITS TO THAT CLINIC.
SECONARY MARKET AREAS INCLUDE OTHER NEIGHBORHOODS WHERE 20% OR
MORE OF EACH NEIGHBORHOOD POPULATION USES THE ASSOCIATED CLINIC.

PRIMARY AND SECONDARY MARKET AREAS*
FOR PRIVATE CLINICS

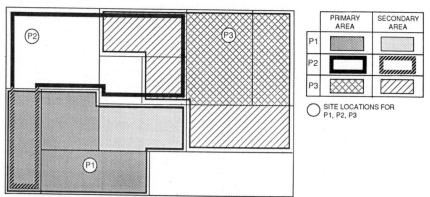

* PRIMARY MARKET AREAS INCLUDE ONLY THOSE NEIGHBORHOODS (a) IN WHICH
25% OR MORE OF THE POPULATION USES THE ASSOCIATED CLINIC AND (b) WHERE
THE NEIGHBORHOODS COLLECTIVELY ACCOUNT FOR OVER 50% OF THE VISITS TO
THAT CLINIC. SECONARY MARKET AREAS INCLUDE OTHER NEIGHBORHOODS WHERE
20% OR MORE OF EACH NEIGHBORHOOD POPULATION USES THE ASSOCIATED CLINIC.

FIGURE 3.19 PATTERNS OF DISPERSAL (**Explanation**: The analyst may wish to differentiate market areas or catchments into a set of nested or hierarchical areas around a focal point. This illustration uses the data from Figure 3.18 to summarize a relatively complex pattern of data. It requires two maps—one would be overly confusing.)

points and overlaps as well as gaps. The summary conveys only the most relevant items, with the full background information shown on other maps.

Multiple maps are particularly useful in issues concerning the interrelationships among data. If several variables can be used in different ways to measure the need for a new social program, the program need could be implied by a very high value for one variable or indicator, a slightly lower value for

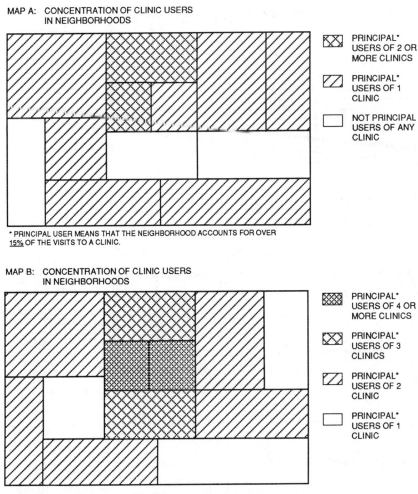

MAP A: CONCENTRATION OF CLINIC USERS
IN NEIGHBORHOODS

PRINCIPAL*
USERS OF 2 OR
MORE CLINICS

PRINCIPAL*
USERS OF 1
CLINIC

NOT PRINCIPAL
USERS OF ANY
CLINIC

* PRINCIPAL USER MEANS THAT THE NEIGHBORHOOD ACCOUNTS FOR OVER
15% OF THE VISITS TO A CLINIC.

MAP B: CONCENTRATION OF CLINIC USERS
IN NEIGHBORHOODS

PRINCIPAL*
USERS OF 4 OR
MORE CLINICS

PRINCIPAL*
USERS OF 3
CLINICS

PRINCIPAL*
USERS OF 2
CLINIC

PRINCIPAL*
USERS OF 1
CLINIC

* PRINCIPAL USER MEANS THAT THE NEIGHBORHOOD ACCOUNTS FOR OVER
5% OF THE VISITS TO A CLINIC.

FIGURE 3.20   PATTERNS OF CONCENTRATION   (**Explanation**: The same data that are used to demonstrate patterns of dispersal can also be used to show concentrations of activity. There are many ways to define and measure such issues. These are two variations of the same issue based on the data of Figure 3.18.)

some pairs of indicators, or just average values for all indicators. Several maps could be prepared, each using different definitions of need and combinations of indicators. One map might be a conservative estimate of need and another a more liberal version. Different maps could reflect different theories, models, or opinions about measuring need (Figure 3.23), and also reflect varying

MAP A: PRIMARY MARKET AREAS * FOR HIGH
         VOLUME CLINICS

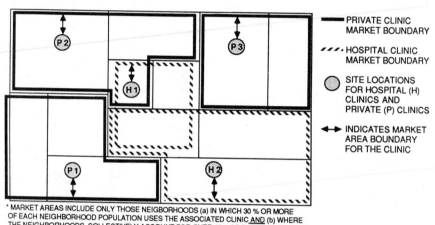

■ PRIVATE CLINIC
MARKET BOUNDARY

▰▰▰ HOSPITAL CLINIC
MARKET BOUNDARY

⬤ SITE LOCATIONS
FOR HOSPITAL (H)
CLINICS AND
PRIVATE (P) CLINICS

◄► INDICATES MARKET
AREA BOUNDARY
FOR THE CLINIC

* MARKET AREAS INCLUDE ONLY THOSE NEIGBORHOODS (a) IN WHICH 30 % OR MORE
OF EACH NEIGHBORHOOD POPULATION USES THE ASSOCIATED CLINIC AND (b) WHERE
THE NEIGHBORHOODS COLLECTIVELY ACCOUNT FOR OVER 40% OF THE VISITS TO THAT
CLINIC

MAP B: PRIMARY MARKET AREAS* FOR HIGH
         VOLUME CLINICS

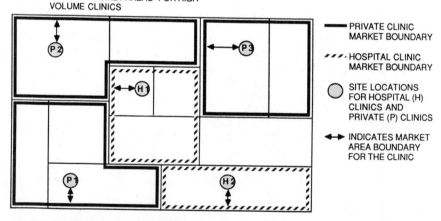

■ PRIVATE CLINIC
MARKET BOUNDARY

▰▰▰ HOSPITAL CLINIC
MARKET BOUNDARY

⬤ SITE LOCATIONS
FOR HOSPITAL (H)
CLINICS AND
PRIVATE (P) CLINICS

◄► INDICATES MARKET
AREA BOUNDARY
FOR THE CLINIC

* MARKET AREAS INCLUDE ONLY THOSE NEIGBORHOODS (a) IN WHICH
35 % OR MORE OF EACH NEIGHBORHOOD POPULATION USES THE
ASSOCIATED CLINIC AND (b) WHERE THE NEIGHBORHOODS COLLECTIVELY
ACCOUNT FOR OVER 25% OF THE VISITS TO THAT CLINIC

FIGURE 3.21 PATTERN SENSITIVITY (**Explanation**: In the preceding set of maps (Figures 3.17 to 3.20) the patterns portrayed are especially sensitive to variations in operational definitions. Consequently, the analyst may wish to experiment with different options for such definitions. Maps A and B are two different alternatives to Map A in Figure 3.18 that present different patterns by slightly modifying an operational definition.)

MAP A: SUMMARY OF RESIDENTIAL DEVELOPMENT ISSUES

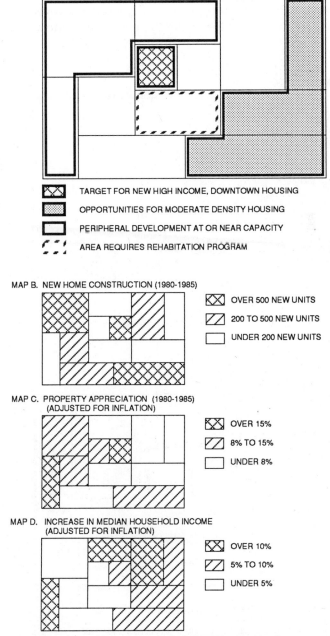

TARGET FOR NEW HIGH INCOME, DOWNTOWN HOUSING

OPPORTUNITIES FOR MODERATE DENSITY HOUSING

PERIPHERAL DEVELOPMENT AT OR NEAR CAPACITY

AREA REQUIRES REHABILITATION PROGRAM

MAP B. NEW HOME CONSTRUCTION (1980-1985)

OVER 500 NEW UNITS

200 TO 500 NEW UNITS

UNDER 200 NEW UNITS

MAP C. PROPERTY APPRECIATION (1980-1985)
        (ADJUSTED FOR INFLATION)

OVER 15%

8% TO 15%

UNDER 8%

MAP D. INCREASE IN MEDIAN HOUSEHOLD INCOME
        (ADJUSTED FOR INFLATION)

OVER 10%

5% TO 10%

UNDER 5%

FIGURE 3.22 SUMMARIZING MULTIPLE VARIABLES (**Explanation:**
One larger map can be used to summarize relevant data from a series
of related maps and to add subjective or qualitative interpretations).

audience perceptions of need. The outcome is that the same data sets can be combined in different ways to show varying geographic patterns.

Multiple maps frequently are used to show changes over time and may, for example, show such changes in the size of subareas with a particular level

MAP A: TARGETING NEW HOUSING IN ECONOMICALLY
        IMPROVING NEIGHBORHOODS

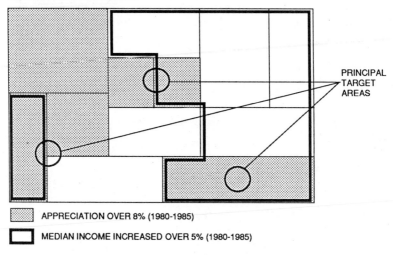

PRINCIPAL
TARGET
AREAS

APPRECIATION OVER 8% (1980-1985)

MEDIAN INCOME INCREASED OVER 5% (1980-1985)

MAP B: TARGETING NEW HOUSING IN MARGINAL
        ECONOMIC AREAS

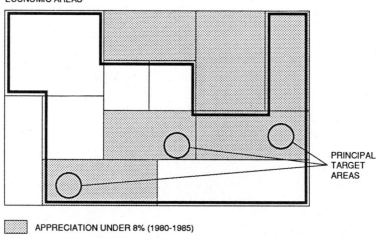

PRINCIPAL
TARGET
AREAS

APPRECIATION UNDER 8% (1980-1985)

MEDIAN INCOME INCREASED 10% OR LESS (1980-1985)

FIGURE 3.23 MAPPING DATA INTO A CONCEPT (*Explanation:* Multiple variables and issues can confuse audiences. Consequently the analyst can use different theories or concepts to link together related data. These two examples are based on the data from Figure 3.22.)

of income, health, education, or other socioeconomic trait (Figure 3.24). Multiple maps also may show temporal changes in the level of a variable in the same size area (Figure 3.25). A common problem arises when the boundaries used for data collection in one time period do not match the boundaries used in the following time period. For example, census tracts, like political boundaries for local governments, are prone to change. Even when boundaries stay the same, variables may not have been measured in precisely the same way in two different time periods. Unfortunately, there is no common remedy. If these difficulties create significant doubts as to the validity or reliability of

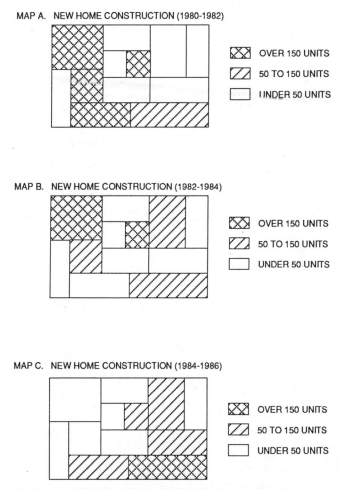

MAP A. NEW HOME CONSTRUCTION (1980-1982)

OVER 150 UNITS
50 TO 150 UNITS
UNDER 50 UNITS

MAP B. NEW HOME CONSTRUCTION (1982-1984)

OVER 150 UNITS
50 TO 150 UNITS
UNDER 50 UNITS

MAP C. NEW HOME CONSTRUCTION (1984-1986)

OVER 150 UNITS
50 TO 150 UNITS
UNDER 50 UNITS

FIGURE 3.24 AREAS CHANGING OVER TIME (**Explanation**: One way to show temporal changes is to portray the same variable at a few points in time. These maps are related to issues shown in Figure 3.22).

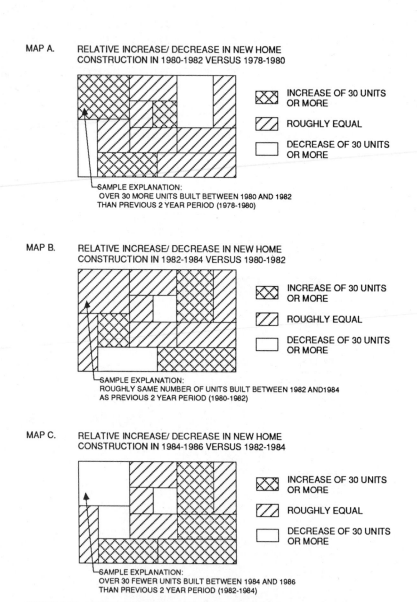

MAP A. RELATIVE INCREASE/ DECREASE IN NEW HOME
CONSTRUCTION IN 1980-1982 VERSUS 1978-1980

☒ INCREASE OF 30 UNITS OR MORE

▨ ROUGHLY EQUAL

☐ DECREASE OF 30 UNITS OR MORE

SAMPLE EXPLANATION:
OVER 30 MORE UNITS BUILT BETWEEN 1980 AND 1982
THAN PREVIOUS 2 YEAR PERIOD (1978-1980)

MAP B. RELATIVE INCREASE/ DECREASE IN NEW HOME
CONSTRUCTION IN 1982-1984 VERSUS 1980-1982

☒ INCREASE OF 30 UNITS OR MORE

▨ ROUGHLY EQUAL

☐ DECREASE OF 30 UNITS OR MORE

SAMPLE EXPLANATION:
ROUGHLY SAME NUMBER OF UNITS BUILT BETWEEN 1982 AND1984
AS PREVIOUS 2 YEAR PERIOD (1980-1982)

MAP C. RELATIVE INCREASE/ DECREASE IN NEW HOME
CONSTRUCTION IN 1984-1986 VERSUS 1982-1984

☒ INCREASE OF 30 UNITS OR MORE

▨ ROUGHLY EQUAL

☐ DECREASE OF 30 UNITS OR MORE

SAMPLE EXPLANATION:
OVER 30 FEWER UNITS BUILT BETWEEN 1984 AND 1986
THAN PREVIOUS 2 YEAR PERIOD (1982-1984)

FIGURE 3.25 TIME CHANGES WITHIN AREAS (**Explanation**: Temporal changes
are frequently shown by keeping the boundaries of areas constant and
portraying the change over time within those given boundaries. For example,
the same hypothetical data set used for Figure 3.24 could also be used for these
maps.)

conclusions, this should be portrayed. When a subarea with a particular incidence of poor housing conditions has increased by 10 or 30 percent (depending on the way data are interpreted), it may be more effective to show this as a range rather than as a precise estimate.

Multiple maps also may be needed if there are limited graphic media to show different issues. Inexpensive presentations may be limited to showing areas only in black and white, or with only two or three types of shading. In these situations, it may be far more effective to show only one variable or issue per map. Thus, the same tone can be repeated on several maps, each indicating a different issue.

Sometimes, it may be useful to convert all variables to binomial distributions, where each subarea either has a particular characteristic (shaded) or does not have that characteristic (not shaded). For example, one map may show high-income areas versus all others, a second map may show only middle-income areas versus all others, a third may show only low-income areas, and a fourth may concentrate solely on poverty (Figure 3.26). The same tone can be used on each map to show the corresponding income level. This may be easier and less confusing than attempting to show all of the information on one map with four tones that are not easily reproduced or that are not visually compatible.

RATIOS, INDICES, AND COMBINED MEASURES

There is extensive literature available on the types of ratios, indices, and other combined measures used to map variables (see Appendix for references). For example, there are economic measures for *specialization, localization,* and *concentration.* Relative increases/decreases over time can be expressed as indices and shown on one map (Figure 3.27).

A common statistical procedure is to observe the values of one set of variables for many different geographic areas, and then to use multivariate statistical techniques (such as multiple regression or factor analysis) to construct a quantitative model. In this way, several socioeconomic variables can be combined in a single measure. For example, the analyst can construct an abstract measure for health risks, neighborhood quality, or economic potential (Figure 3.28).

These techniques are statistically complex and generate figures that are difficult to comprehend without detailed explanation. At the very least, they should be accompanied by illustrations of their meaning and interpretations of a few select cases.

When using abstract or complex statistics, it is necessary to show the audiences the meaning of the relative values. Any measuring technique that is applied to a series of geographic areas will result in a range of values, although the relative differences in value may not be significant to the issue and the

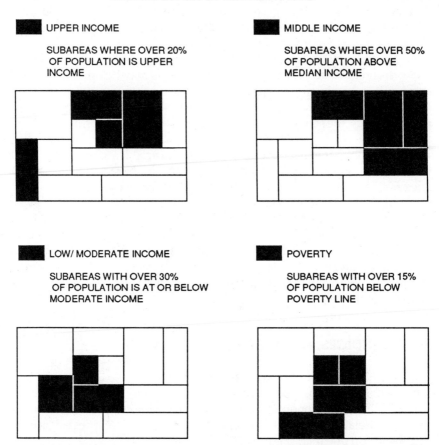

INCOME LEVELS BY NEIGHBORHOOD

UPPER INCOME

SUBAREAS WHERE OVER 20%
OF POPULATION IS UPPER
INCOME

MIDDLE INCOME

SUBAREAS WHERE OVER 50%
OF POPULATION ABOVE
MEDIAN INCOME

LOW/ MODERATE INCOME

SUBAREAS WITH OVER 30%
OF POPULATION IS AT OR BELOW
MODERATE INCOME

POVERTY

SUBAREAS WITH OVER 15%
OF POPULATION BELOW
POVERTY LINE

FIGURE 3.26   MAPPING BINOMIAL DATA   (**Explanation**: In some instances the analyst must use limited graphic media for producing or disseminating results. In other situations the analyst may have to present several overlapping, complex issues. In these instances it may be useful to present a series of maps, each one containing a single dichotimized issue.)

audience (even if they are statistically significant). For instance, an analyst could recode the values of an abstract indicator into five ordinal categories ranging from very high to very low, just to display the geographic pattern of the data (Figure 3.28). The analyst should first be confident that a five-point scale shows differences that are relevant to the issues and the audience and not just to the structure of the data. Applying a scale simply to show the range of the data is inappropriate. In another example, a five-point scale may be used to show data related to the need for a program, where the top three data categories are clear indicators of program need and the distinctions among

RELATIVE PERCENTAGE OF NEW HOME CONSTRUCTION IN EACH
NEIGHBORHOOD 1982-1984 VERSUS 1980-1982

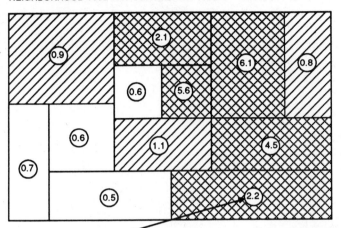

SAMPLE CALCULATION:

NEIGHBORHOOD CONSTRUCTION 1982-84 = $\frac{200}{1120}$ = 18%
ALL CONSTRUCTION 1982-84 = 1120

NEIGHBORHOOD CONSTRUCTION 1980-82 = $\frac{100}{1250}$ = 8%
ALL CONSTRUCTION 1980-82 = 1250

RELATIVE PERCENTAGE = (18%) / (8%) = 2.2

NEIGHBORHOOD WITH SIGNIFICANTLY
HIGHER % OF TOTAL HOUSING CONSTRUCTION
IN RECENT VERSUS PREVIOUS PERIOD

NEIGHBORHOOD WITH ROUGHLY
SAME % OF CONSTRUCTION IN RECENT
VERSUS PREVIOUS PERIOD

NEIGHBORHOOD WITH SIGNIFICANTLY
LOWER % OF CONSTRUCTION IN
RECENT VERSUS PREVIOUS PERIOD

RELATIVE PERCENTAGE OF NEW HOME CONSTRUCTION IN EACH
NEIGHBORHOOD 1984-1986 VERSUS 1982-1984

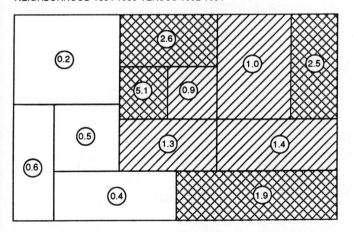

FIGURE 3.27   INDICES ON MAPS   (**Explanation**: Frequently analysts use an index ratio, coefficient, or similar measure to map a spatial pattern. These two examples follow the issues discussed in Figures 3.22 to 3.25.)

categories reflect only minor differences. It may be more effective to use only two codes instead of five to designate the subareas that, in the opinion of the analyst, represent a relevant difference. If the analyst cannot judge what will be meaningful to the audience and the issues being discussed, the value of the whole presentation may be lessened.

INDEX OF PATTERN OF RESIDENTIAL QUALITY *

MAP A. (BASIC DATA)

.88		.63	.15	.28	.72			
					.35	.76	.91	
				.18				
.45	.69	.51	.62	.70	.39	.35	.33	
			.58	.52	.22	.20		
.50	.42	.33	.20	.24		.35	.73	
					.85			
.65		.66		.90		.72		.78

* BASED ON COMBINED MEASUREMENTS OF: HOUSING OWNERSHIP, AGE OF HOUSING,
AVAILABLE PUBLIC AMENITIES, MEDIAN INCOME, LOCAL SCHOOLS, CRIME RATE, RATE
OF MORBIDITY, UNEMPLOYMENT (THE INDEX RANGES FROM 0 TO 1.00)

MAP B. (VISUAL PATTERN)

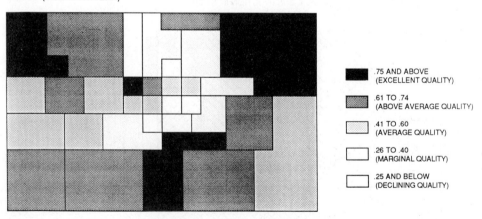

■ .75 AND ABOVE
(EXCELLENT QUALITY)

▨ .61 TO .74
(ABOVE AVERAGE QUALITY)

▤ .41 TO .60
(AVERAGE QUALITY)

□ .26 TO .40
(MARGINAL QUALITY)

□ .25 AND BELOW
(DECLINING QUALITY)

FIGURE 3.28 TRANSLATING ABSTRACTIONS ON MAPS (***Explanation***: If the analyst uses too abstract a measure then it is appropriate to present some codified interpretation for the audience. Map A shows only abstract data. Map B shows the same data, interpreted for a general audience.)

4

Statistics For Decisions:
Benefits Versus Costs

Benefit-cost analysis is a decision-making technique based on an elementary concept. The advantages and disadvantages of an action are compared and, if the comparison is favorable, the action is taken. However, this premise becomes more complicated as soon as the inherent underlying assumptions are examined.

The first assumption of some benefit-cost analyses is that the pros and cons can be easily translated into comparable quantitative or monetary terms. In practice, however, there is frequently a need to compare apples and oranges or, more accurately, to compare a variety of qualitative outcomes that are not easily measured.

A second frequent assumption is that a benefit-cost analysis includes all the actions worthy of consideration, but in every instance there is always the possibility that a hitherto undiscovered action would produce a superior outcome; that is, a more favorable weighing of benefits versus costs. When two actions are being compared, there is always the possibility of a third; when three are being studied, a fourth; and so on.

A third assumption is that the full range of benefits and costs, even if they are nonquantifiable, are included. In practice, however, a particular cost or benefit will lead to further benefits and costs. For example, the benefits of an economic development program might include the creation of new jobs, but these, in turn, create new outcomes such as decreased unemployment benefits, increased population growth, and new demands for public services. Therefore, it is sometimes difficult to decide where to end the trail of costs and benefits.

A fourth assumption is that there will be a clear criterion for deciding which package of benefits and costs is superior. One action might cost $200,000 and have benefits of $500,000. Another action might cost $1,000,000 and have benefits of $1,400,000. It is a debatable issue as to which action is superior. Costs may be subtracted from the benefits and the absolute difference in value examined, or the percentage of increase, or rate of return, of benefits versus costs analyzed. Each approach will lead to a different conclusion. Dealing with these assumptions and their ensuing logical difficulties is the crux of the problem in presenting a benefit-cost analysis effectively.

NONQUANTIFIABLE COSTS AND BENEFITS

The problem of presenting qualitative as well as quantitative data is one of the most frequent pitfalls of presentations. If all of the information is measurable and comparable, there is no problem, but these situations are rare. Furthermore, many benefit-cost analyses are challenged at some point by the audience with the observation that important nonquantifiable or noncomparable data have been ignored. Too often, this is a legitimate challenge, as analysts sometimes ignore these potential problems in an effort to generate an internally consistent statistical argument with a clear answer. It is difficult for many analysts to accommodate qualitative issues, as it often leads to less clear, less consistent arguments, even if they are more relevant. The following subsections give some direction in approaching this problem.

Deemphasizing Nonquantifiable and Noncomparable Values

A direct approach to presenting qualitative and noncomparable items is to try to have your cake and eat it too. One part of the presentation, in table form, specifies the quantifiable and comparable values for costs and benefits. A secondary table, or a subcomponent of the principal table, lists the other relevant values in a visually deemphasized manner (Figure 4.1). This approach allows the presentor to focus on an internally consistent benefit cost analysis, while simultaneously informing the audience that other issues have been recognized but are considered of secondary importance. Obviously, this technique will be effective *only* when there is a strong argument supporting the lessened importance of the secondary information.

Cost Effectiveness

Another technique in presenting qualitative and noncomparable values involves showing some quantitative information, but asserting that it is insufficient for making a decision. Rather, the decision must be arrived at through subjective evaluation, which rests on both qualitative and quantitative information.

DIAGRAM A. COSTS AND BENEFITS FOR PUBLIC HEALTH PROGRAM OPTIONS

	PROGRAM X HOME HEALTH CARE FOR THE ELDERLY	PROGRAM Y IMMUNIZATION PROGRAM FOR SCHOOL CHILDREN
COSTS (IN MILLIONS OF DOLLARS)		
A. STAFF	2.6	4.1
B. FACILITY RENTAL/ FINANCIAL COSTS	0.8	0.5
C. SUPPLIES AND EXPENSES	0.5	1.0
D. OVERHEAD	1.0	1.3
TOTAL ANNUAL COST	4.9	6.9
BENEFITS (IN MILLIONS OF DOLLARS)		
A. SAVIINGS IN HOSPITAL STAFF AND TREATMENT COSTS	1.5	2.9
B. SAVINGS IN OPERATIONAL COSTS FOR EXISTING FACILITIES	2.2	2.0
TOTAL	0.7	4.9

OTHER ISSUES: 1) NUMBER OF PEOPLE/ FAMILIES SERVED
2) INCREASE IN LIFE EXPECTANCY
3) SOCIAL RESPONSIBILITY

DIAGRAM B. COSTS AND BENEFITS FOR PUBLIC HEALTH PROGRAM OPTIONS

	PROGRAM X HOME HEALTH CARE FOR THE ELDERLY	PROGRAM Y IMMUNIZATION PROGRAM FOR SCHOOL CHILDREN
COSTS (IN MILLIONS OF DOLLARS)		
A. STAFF	2.6	4.1
B. FACILITY RENTAL/ FINANCIAL COSTS	0.8	0.5
C. SUPPLIES AND EXPENSES	0.5	1.0
D. OVERHEAD	1.0	1.3
TOTAL ANNUAL COST	4.9	6.9
EFFECTS/ COST SAVINGS (IN MILLIONS OF DOLLARS)		
A. SAVIINGS IN HOSPITAL STAFF AND TREATMENT COSTS	1.5	2.9
B. SAVINGS IN OPERATIONAL COSTS FOR EXISTING FACILITIES	2.2	2.0
TOTAL	3.7	4.9
SOCIAL BENEFITS		
A. NUMBER OF PEOPLE SERVED	15,000	180,000
B. AVERAGE INCREASE IN LIFE EXPECTANCY PER PATIENT (IN YEARS)	0.2	0.1

FIGURE 4.1 COMPARING INCOMPARABLES (*Explanation:* When one presents costs and benefits, the most typical problems involve comparing items that are qualitatively different or that are not quantifiable. Diagrams A and B illustrate two ways to do this. Diagram A simply deemphasizes the noncomparable values in a footnote. Diagram B includes noncomparable values and, as a result, makes the decision-making situation more ambiguous. Which approach is more appropriate depends on the nature of the problem and the audience.)

For example, the financial costs of two social programs may be reliably projected and compared, but the ensuing benefits are simply noncomparable and/or qualitative. Program *A* might provide substantial health benefits for a small elderly population, while program *B* provides lesser benefits but to a larger population of infants (Figure 4.1). This is a case in which relevant data can be presented, but will not be enough to make a final program choice. The decision ultimately requires a subjective, and hopefully enlightened, review of qualitative factors.

This technique often is referred to as a *cost-effectiveness analysis*, stemming from the premise that two or more actions with comparable monetary costs can be compared in terms of the noncomparable effects they produce. This is quite normal, although the more common situation is one in which costs as well as benefits contain both comparable numbers and qualitative and noncomparable values.

A more problematic situation arises when radically different qualitative issues must be compared. If one program provides health benefits to one group while an alternative program provides economic development benefits to a different group, the analyst may present lengthy, complex rationales for recommending one or the other program. The argument often rests on finding some common denominator. For instance, both programs could be discussed in the context of improving the overall welfare of the larger community in which they are located (Figure 4.2). The analysis should be tempered with statements regarding the use of subjective judgments, and the presentor might wish to show that judgments by individuals or groups (other than those responsible for the analysis) may be needed before an obvious choice will emerge.

Converting Costs and Benefits: Creating Comparables

It is possible to convert qualitative and noncomparable data into other variables or indices so that all items can be measured and compared. For example, the benefits of two health programs can be converted into an estimate of the number of added years of life that accrue to the people served. This may be taken one step further. Added years of life can be converted into the expected dollar value of added economic productivity of the people being served. There may be still other monetary measures of an added year of life based, perhaps, on life insurance statistics. Once the data are converted to comparable quantities, the benefit-cost analysis is simplified.

Such computations are complex and may be questionable, although that is not at issue here. If the analyst does select this approach, how should it be presented? Most audiences may not understand or appreciate such conversion techniques, so the presentor should create a series of tables, each table moving one step further in the conversion process (Figure 4.3). This allows the audience to understand the procedures and the assumptions, but does not make the presentor appear to be hiding something or mystifying the process.

DIAGRAM A: COSTS AND ORIGINAL NON-COMPARABLE BENEFITS

	OPTION A JOB CREATION & ECONOMIC DEVELOPMENT PROGRAM	OPTION B INDUSTRIAL HEALTH CARE PROGRAM
COSTS: ANNUAL COST	$1,800,000	$700,000
BENEFITS: ECONOMIC BENEFITS	A. 100 NEW JOBS/ YEAR B. GROWTH OF ECONOMIC BASE C. REDUCED UNEMPLOYMENT RATE	A. REDUCED ABSENTEEISM B. INCREASED WORKER PRODUCTIVITY
SOCIAL BENEFITS	A. IMPROVEMENT IN POOR NEIGHBOR- HOODS B. IMPROVED BUSINESS MORALE	A. IMPROVED QUALITY OF LIFE B. IMPROVED JOB SATISFACTION

DIAGRAM B: COSTS AND MODIFIED NON-COMPARABLE BENEFITS

	OPTION A JOB CREATION & ECONOMIC DEVELOPMENT PROGRAM	OPTION B INDUSTRIAL HEALTH CARE PROGRAM
COSTS: ANNUAL COST	$1,800,000	$700,000
BENEFITS: NEW JOBS REDUCED UNEMPLOYMENT REDUCED ABSENTEEISM INCREASED PRODUCTIVITY	 2,000,000 500,000 0 0	 0 200,000 500,000 300,000
TOTAL	$2,500,000	$1,000,000

OTHER ISSUES: NEIGHBORHOOD IMPROVEMENT
IMPROVED BUSINESS MORALE
JOB SATISFACTION

FIGURE 4.2 MODIFYING NONCOMPARABLES: POINTS AND PATTERNS (*Explanation*: In some cases it is possible to convert qualitatively different outcomes into a comparable, sometimes quantitative measure. Diagram A shows two qualitatively different programs that can be modified into comparable measures, as is shown in Diagram B. In this instance, Diagram B makes the comparison easier, but deemphasizes qualitative differences, as is shown in Diagram A, which may be more important to the audience.)

Another technique frequently used is the conversion of future monetary costs and benefits into *present values*. These techniques sometimes are called *discounting* or calculating a *net present value*. One program may require a cost of $100,000 to be spent in the current year, while an alternative program may require a cost of $150,000 five years from now. Can these programs be compared given inflation rates, and what are the monetary benefits (in today's dollars) for programs that generate varying amounts of revenue for each year over the next two decades? Computational formulas for answering these questions are available, but the presentor still faces the problem that most audiences are unfamiliar with such calculations and their significance.

An audience may be willing to accept the presentor's assertion that the computation of present values converts dollar amounts for different years into comparable amounts for the present. However, there are many assumptions about interest rates, opportunity costs, and other economic uncertainties that are inherent in these calculations. Audience awareness of these hidden assumptions as well as their unfamiliarity with the entire concept can create a presentation problem. The use of a computation example — on a separate page, in an appendix, or to the side of the principal table — is an effective way to explain the technique.

Converting Costs and Benefits: Successive Comparisons

The final conversion technique to be considered is a complex process requiring successive trial comparisons between items, where each comparison can be thought of as an equation. The analyst evaluates both sides of the equation to see if they are balanced or of equal value. If they are not, one side of the equation is changed until both sides are equal. This is called the *point of indifference*. After many successive comparisons and indifference points, the analyst can develop a rough approximation of the relative value of qualitative, noncomparable items. The process is painstaking, but can produce relevant answers.

For example, when analyzing the two health programs previously noted, the question could be asked: Given the *benefits*, do you prefer program X or program Y? If the answer is program X, the next question might call for a choice between program X versus program Y plus a cost savings of $500,000. If the answer is still program X, the next comparison would increase the hypothetical added cost savings for program Y. This would continue until the person or group making the hypothetical choice could no longer decide. At this point, the two sets of outcomes are considered roughly equal. The hypothetical cost savings at this point of indifference can then be used as a factor in a formal, comprehensive comparison of costs and benefits. If the two programs are considered equal, when program Y also includes a cost savings of $1,000,000, *and* if program Y in fact will cost $2,000,000 less, the analysis would favor choosing program Y (Figure 4.4). Unfortunately, this example is a simplified case. In practice, such conversion techniques often involve hundreds of

DIAGRAM A. COSTS AND BENEFITS FOR PUBLIC HEALTH PROGRAM OPTIONS

	PROGRAM X HOME HEALTH CARE FOR THE ELDERLY	PROGRAM Y IMMUNIZATION PROGRAM FOR SCHOOL CHILDREN
TABLE A. PUBLIC COSTS COSTS (IN MILLIONS OF DOLLARS)		
A. STAFF	2.6	4.1
B. FACILITY RENTAL/ FINANCIAL COSTS	0.8	0.5
C. SUPPLIES AND EXPENSES	0.5	1.0
D. OVERHEAD	1.0	1.3
TOTAL ANNUAL COST	4.9	6.9
TABLE B. PUBLIC SAVINGS BENEFITS (IN MILLIONS OF DOLLARS)		
A. SAVINGS IN HOSPITAL STAFF AND TREATMENT COSTS	1.5	2.9
B. SAVINGS IN OPERATIONAL COSTS FOR EXISTING FACILITIES	2.2	0.0
TOTAL	3.7	4.9
TABLE C: SOCIAL BENEFITS		
A. NUMBER OF PEOPLE SERVED	15,000	180,000
B. AVERAGE INCREASE IN LIFE EXPECTANCY PER PATIENT (IN YEARS)	0.2	0.1
TABLE D: ECONOMIC VALUE OF SOCIAL BENEFITS		
TOTAL ADDED YEARS OF LIFE BY AGE GROUP		
0-20	0	2,700
20-65	600	4,500
65+	2,400	10,800
VALUE OF ADDED LIFE YEARS MEASURED IN EXPECTED ECONOMIC PRODUCTIVITY		
0-20 ($5,000/ Yr. x # of years)	$0	$13.5 million
20-65 ($30,000/ Yr. x # of years)	$18 million	$135 million
65+ ($20,000/ Yr. x # of years)	$48 million	$216 million
TOTAL ECONOMIC BENEFIT	$66 million	$364.5 million

FIGURE 4.3 CREATING COMPARABLES (**Explanation**: The following sequence of tables shows how costs and benefits, similar to Figure 4.1, can be converted into comparable figures. The goal is to create comparable dollar measurements for the same variable. It should also be noted that the cost savings noted in Table B may be the direct economic benefit to the government that sponsors the health program and that the economic value of the several benefits shown in Table D may or may not be the important economic issue for the decision makers.)

subjective judgments, which are further complicated by varying estimates as to the probabilities and values of the outcomes.

Another technique, based on similar principles, is the creation of a new index for measuring benefits and costs. For example, points could be assigned to each component of the costs and benefits, such as points for capital costs, points for annual operating costs, points for number of people served, points for geographic location, and so on. Taking this idea one step further, types of component or factor could be weighted numerically before points are assigned, so *capital costs* could be weighted 3.5, *people served* could be weighted 2.0, and so on. The points for each factor are multiplied by the weight, and the sum total becomes the index, which purportedly measures the overall value of the program. The problem is how to present such a computational process in a way that is clear and convincing to the audience.

At the very least, the presenter must show how the points and weights were established and compiled. More importantly, the sensitivity of the analytic conclusion to minor fluctuations in subjective judgments, point allocations, and weighting factors should be illustrated (Figure 4.5). There is no simple way to do this without risking distortion or misleading the audience, although the most effective approach would be to show the audience a summary table or chart with the "bottom line" and then a series of tables indicating the steps in the process and the sensitivity of the evaluation to minor changes. Ordinarily, this could be accomplished with three to four visual displays. The technique of combining nonquantifiable and noncomparable items is similar to techniques elaborated in the next chapter in the section entitled *Ranking and Weighting*.

SETS OF COSTS AND BENEFITS

In addition to qualitative and noncomparable items, another significant problem concerns the range and variety of costs and benefits. Any action is going to create ripples of benefits and costs that, in effect, never end. For example, the benefits of a health program accrue not just to the people who directly receive health services, but to their families, employers, and health insurance companies. These are secondary or second-order benefits that, in turn, create tertiary or third-order benefits for other parties and ultimately to society as a whole. This can be further illustrated by looking at a case in which an economic development program creates new jobs. The persons receiving income from these jobs spend their money on goods and services, so there is a continuing ripple of economic benefit to the remainder of the community.

A cost for one group may be a benefit for another. A new social program is a cost to the public agency that provides the funding and ultimately a cost to the taxpayers who support that government. However, it can be a benefit to all those who will be employed in the new program and, in turn, the community in which they spend their income. In some cases, there may be statistical

DIAGRAM A. COSTS AND BENEFITS FOR PUBLIC
HEALTH PROGRAM OPTIONS

	PROGRAM X HOME HEALTH CARE FOR THE ELDERLY	PROGRAM Y IMMUNIZATION PROGRAM FOR SCHOOL CHILDREN
EFFECTS -SOCIAL BENEFITS NUMBER OF PEOPLE SERVED -AVERAGE INCREASE IN LIFE EXPECTANCY	15,000 0.2	180,000 0.1
BENEFITS SAVINGS IN HOSPITAL, STAFF & TREATMENT COSTS -SAVINGS IN OPERATIONAL COSTS FOR EXISTING FACILITIES	$1,500,000 $2,200,000	$2,900,000 $2,000,000
EXPECTED PREFERENCE IF COSTS ARE EQUAL		✔
EXPECTED PREFERENCE IF COST OF PROGRAM X IS LESS THAN THE COST OF PROGRAM Y BY: $100,000		✔
$250,000		✔
$500,000		✔
INDIFFERENCE POINT $1,000,000 $750,000		✔
$1,000,000	✔	
$1,250,000	✔	
ACTUAL COST DIFFERENCE: PROGRAM X IS $2,000,000 LESS THAN PROGRAM Y	ACTUAL COST $4,900,000	ACTUAL COST $6,900,000
RECOMMENDED CHOICE:	✔	

FIGURE 4.4 COMPARISON AT THE INDIFFERENCE POINT
(*Explanation*: The goal of this technique is to demonstrate that there
is some point at which the positive benefits or effects of two options
can be considered roughly equal, depending on the relative
difference in their costs. It is like adding weights to one side of a
scale until both sides are balanced. After the balance point (or point
of indifference) is estimated, this can be compared to the actual
difference in the costs. The results of this line of reasoning can be
presented as is shown here.)

TABLE A. EVALUATION OF HEALTH PROGRAM OPTIONS

	RELATIVE WEIGHT OF FACTOR *	PROGRAM X HOME HEALTH CARE FOR THE ELDERLY	PROGRAM Y IMMUNIZATION PROGRAM FOR SCHOOL CHILDREN
COST OF CAPITAL CONSTRUCTION	3.5	40	60
OTHER COSTS	5.0	65	35
PUBLIC COST SAVINGS	3.0	45	55
NUMBER OF PEOPLE SERVED	2.0	20	80
AVERAGE INCREASE IN LIFE EXPECTANCY	2.5	35	65
TOTAL SCORE (SUM OF WEIGHT X POINTS)		727.50	872.50

* BASED ON SUBJECTIVE ESTIMATES FROM TWELVE MEMBERS
  OF APPOINTED DECISION REVIEW PANEL

TABLE B. POTENTIAL FLUCTUATIONS IN FINAL SCORES

	RELATIVE WEIGHT OF FACTOR *		PROGRAM X		PROGRAM Y	
	HIGH	LOW	HIGH	LOW	HIGH	LOW
COST OF CAPITAL CONSTRUCTION	3.8	3.3	45	30	70	55
OTHER COSTS	5.1	4.8	80	55	45	20
PUBLIC COST SAVINGS	3.2	2.6	55	15	85	45
NUMBER OF PEOPLE SERVED	3.0	1.5	30	15	85	70
AVERAGE INCREASE IN LIFE EXPECTANCY	3.5	1.5	70	10	90	30
TOTAL SCORE FOR EACH OF TWELVE PANEL MEMBERS			810	350	1250	680
NUMBER OF TIMES PROGRAM RECEIVED HIGHEST SCORE (OUT OF 12 PANEL MEMBERS)			3		9	

FIGURE 4.5   CREATING A WEIGHTED POINT INDEX   (***Explanation**:* Any quantitative or qualitative values can be converted to a subjective point index. Typically this is done by asking a panel of experts or decision makers to review all of the items and to assign points and weights. In the following example Table A shows the summary of a panel's subjective judgments, and Table B shows the sensitivity of these judgments relative to variations in weights and points.)

procedures for computing the trail of costs, as in the case of methods for calculating economic multiplier effects for increases in the economic base of a community. More often than not, such computational algorithms are nonexistent, and the analyst has to decide how to limit and group the various costs and benefits to create the most effective argument.

Grouping by Orders of Costs and Benefits

The most common analytic practice is to group costs and benefits as first-order, second-order, third-order, and so on. First-order, or primary benefits and costs, usually are those that are directly relevant to the decision. Secondary benefits and costs are those created by the primary benefits and cost. They are typically indirect and often have some relevance to the decision. Third-order, or tertiary benefits and costs, are created by the secondary elements, and presumably have even less relevance to the decision. Carrying the analysis and its presentation beyond the tertiary level is likely to be too complex, so if there are further benefits and costs they should, perhaps, be combined in the last category under a heading such as *third-order and other benefits and costs.*

The sets of costs and benefits should be presented as a series of tables, with the greatest visual emphasis given to the primary category (Figure 4.6). However, several presentation problems can arise. The analysis of primary benefits and costs may favor action X over action Y, but when secondary benefits and costs are included, the opposite conclusion emerges. It cannot be presumed that the inclusion of secondary and/or tertiary benefits and costs is automatically a superior or more rationale approach, as this will depend on the nature of the problem and the audience being addressed. An effective way to solve this dilemma would be to create two tables, one with only primary benefits and costs the other with both primary and secondary benefits and costs. The audience then does not have to go through the mental exercise of aggregating the data. Of course, visual emphasis should be given to the table that, in the analyst's view, represents the more relevant approach.

A related presentation problem occurs when only some of the secondary or tertiary benefits and costs are relevant. For example, in comparing two health programs, a relevant secondary benefit may be the improved well-being of the family of the person receiving service. On the other hand, the improved economic productivity of the person receiving health services may not, from the viewpoint of the analyst and the audience, be especially relevant. In such circumstances, the analyst should present the relevant primary, secondary, and tertiary benefits and costs as clearly separate from the nonrelevant items in each category. One chart or table should categorize all items, regardless of relevance, according to primary, secondary, and tertiary levels. Another visually dominant chart (or column in the table) should then accumulate only relevant benefits and costs regardless of their primary, secondary, or tertiary nature (Figure 4.6).

Grouping Benefits and Costs by Coalitions

A less systematic approach to grouping benefits and costs is their organization according to different interest groups, constituencies, coalitions, or relevant subdivisions of the audience. For example, a social program might have different benefits and costs for different geographic communities or age groups. Similarly, two economic development programs might have different

TABLE A.

	COSTS		BENEFITS / EFFECTS	
PRIMARY				
PROGRAM Y: IMMUNIZATION PROGRAM FOR SCHOOL CHILDREN	FACILITY	$500,000	SAVED STAFF COSTS	$2,900,000
	STAFF	$3,100,000		
	SUPPLIES	$1,000,000	SAVED OPERATIONAL COSTS	$2,000,000
	TOTAL	$4,600,000	TOTAL	$4,900,000
PROGRAM X: HOME HEALTH CARE FOR THE ELDERLY	FACILITY	$800,000	SAVED STAFF COSTS	$1,500,000
	STAFF	$2,100,000		
	SUPPLIES	$500,000	SAVED OPERATIONAL COSTS	$2,200,000
	TOTAL	$3,400,000	TOTAL	$3,700,000
SECONDARY PROGRAM Y:	ADMINSTRATION	$1,000,000	NUMBER OF PEOPLE SERVED	180,000
	INDIRECT COST	$1,300,000		
PROGRAM X:	ADMINSTRATION	$500,000	NUMBER OF PEOPLE SERVED	15,000
	INDIRECT COST	$1,000,000		
PROGRAM Y:	TRANSIT COSTS	0	INCREASED TAX REVENUE	$8,400,000
			INCREASED PRODUCTIVITY	$364,500,000
PROGRAM X:	TRANSIT COSTS	$400,000	INCREASED TAX REVENUE	$2,300,000
			INCREASED PRODUCTIVITY	$66,000,000

FIGURE 4.6 ORDERING COSTS, BENEFITS, AND EFFECTS (**Explanation:** Many analysts are trained to segregate costs, benefits, and effects into primary, secondary, and tertiary levels. This practice should be tailored to suit the problem and the audience. All of the following tables contain the same data. Table A desegregates items into first, second, and tertiary (third level) issues. Table B combines the first and second levels and deemphasizes the third. Table C shows how the same items can be reorganized in a still more subjective manner to suit presumably specific needs of the analyst and audience. The data match the example shown in Figure 4.1.)

TABLE B.

MAJOR ISSUES:	COSTS		BENEFITS / EFFECTS	
PROGRAM Y: IMMUNIZATION PROGRAM FOR SCHOOL CHILDREN	FACILITY STAFF SUPPLIES ADMINISTRATION INDIRECT COSTS	$500,000 $3,100,000 $1,000,000 $1,000,000 $1,300,000	SAVED STAFF COSTS SAVED OPERATIONAL COSTS # OF PEOPLE SERVED	$2,900,000 $2,000,000 (180,000)
TOTALS		$6,900,000		$4,900,000
PROGRAM X: HOME HEALTH CARE FOR THE ELDERLY	FACILITY STAFF SUPPLIES ADMINISTRATION INDIRECT COSTS	$800,000 $2,100,000 $500,000 $500,000 $1,000,000	SAVED STAFF COSTS SAVED OPERATIONAL COSTS # OF PEOPLE SERVED	$1,500,000 $2,200,000 (15,000)
TOTALS		$4,900,000		$3,700,000
MINOR ISSUES: PROGRAM Y:	TRANSIT COSTS	0	INCREASED TAX REVENUE INCREASED PRODUCTIVITY	$8,400,000 $364,500,000
PROGRAM X:	TRANSIT COSTS	$400,000	INCREASED TAX REVENUE INCREASED PRODUCTIVITY	$2,300,000 $66,000,000

TABLE C.

MAJOR ISSUES:	COSTS		BENEFITS / EFFECTS	
PROGRAM Y: IMMUNIZATION PROGRAM FOR SCHOOL CHILDREN	FACILITY STAFF SUPPLIES ADMINISTRATION TRANSIT COSTS	$500,000 $3,100,000 $1,000,000 $1,000,000 $0	SAVED STAFF COSTS SAVED OPERATIONAL COSTS INCREASED TAX REVENUE	$2,900,000 $2,000,000 $8,400,000
TOTALS		$5,600,000		$4,900,000
PROGRAM X: HOME HEALTH CARE FOR THE ELDERLY	FACILITY STAFF SUPPLIES ADMINISTRATION TRANSIT COSTS	$800,000 $2,100,000 $500,000 $500,000 $1,000,000	SAVED STAFF COSTS SAVED OPERATIONAL COSTS INCREASED TAX REVENUE	$1,500,000 $2,200,000 $2,300,000
TOTALS		$4,900,000		$3,700,000
OTHER VARIABLES AND ISSUES: PROGRAM Y:	INDIRECT COSTS	$1,300,000	NUMBER OF PEOPLE SERVED INCREASED PRODUCTIVITY	180,000 $364,500,000
PROGRAM X:	INDIRECT COSTS	$1,000,000	NUMBER OF PEOPLE SERVED INCREASED PRODUCTIVITY	15,000 $66,000,000

costs and benefits for different segments of the work force (such as union workers, unemployed youths, unemployed heads of families, skilled workers, and unskilled workers), or the members of a decision-making committee may represent different political coalitions with varying needs and aspirations.

Too often, analysts avoid these political realities and attempt to provide a single, narrowly focused analysis. The result usually is that one group or coalition immediately perceives the analysis as favorable, whereas another views it as unfavorable. The debate then shifts to whether the analysis should be accepted, rather than on how the analysis can be used to reach a reasonable compromise.

An effective presentation approach recognizes potentially controversial concerns of the audience and presents the analysis accordingly. The presentation could show the benefits and costs for group A, B, C, and so on. This does not necessarily mean that the labels on each benefit-cost table are *Political Group A* or *Political Group B*, although this may be appropriate when each such group is, for example, a unit of government. Typically, it may be more appropriate to label the tables or charts according to the types of persons each political coalition may represent, such as types of workers, neighborhoods, regions, economic sectors, professional groups, age groups, socioeconomic status, and the like. This does not hide political realities, but rather presents the data so that each coalition can comprehend the basis of the opinions or judgements of the other coalitions (Figure 4.7).

UNSEEN ALTERNATIVES: *DO NOTHING* AND *NEW PROPOSAL*

In the simplest example, a cost-benefit or cost-effectiveness analysis can be used for one action. If benefits exceed costs, the answer seems straightforward—decide in favor of taking the action. In practice, however, the simple case of examining one alternative implies a comparison to at least two other possible courses of action that are *do nothing* and *new proposal*. Choosing the former option rests largely on the merits of the status quo. Choosing the latter options rests on the potential for new ideas and emerging opportunities. Even in the simplest case of reviewing one action, the presentor must be aware of these other two, often hidden, possibilities.

It may be necessary to formalize these two universal options in table form alongside the principal benefit-cost analysis, perhaps where there is only a minor positive difference between benefits versus costs, or where benefits and costs may be uncertain, creating some doubt as to whether benefits will truly exceed costs or vice versa. In both cases, either taking no action or generating new proposed actions may be a wise choice.

The detail with which *do nothing* and *new proposal* options are presented depends on how much consideration they deserve. At one extreme, they may be presented as side notes on a table, perhaps under a heading such as *other considerations*. The audience is then aware that these possibilities have been considered, although they appear to have marginal relevance, and are worth further attention only if there is a serious problem with the other arguments for taking action (Figure 4.8).

At another level, one or both of these options can be presented in a form paralleling other costs and benefits, rather than presented as side notes. Greater visual emphasis should be given to the principal action under consideration, although this may mean that there are some relevant items of data and

COSTS AND BENEFITS BY GROUPS AND PARTICIPATING ORGANIZATIONS

PROGRAM A. JOB CREATION AND ECONOMIC DEVELOPMENT PROGRAM		TOTALS	LOCAL GOVERNMENT	BUSINESS AND INDUSTRY	EMPLOYEES
COSTS	STAFF	$1,000,000	8,000,000	$200,000	$0
	FACILITIES	200,000	200,000	0	0
	OVERHEAD/ ADMINISTRATION	600,000	600,000	0	0
	TOTALS	$1,800,000	$1,600,000	$200,000	$0
BENEFITS	NEW JOBS	$2,000,000	$250,000	$400,000	$1,350,000
	REDUCED UNEMPLOYMENT	500,000	150,000	150,000	200,000
	TOTALS	$2,500,000	$400,000	$2,500,000	$1,550,000

PROGRAM B. INDUSTRIAL HEALTH CARE PROGRAM					
COSTS	STAFF	$150,000	50,000	$100,000	$0
	FACILITIES	300,000	100,000	200,000	0
	OVERHEAD/ ADMINISTRATION	250,000	100,000	150,000	0
	TOTALS	$700,000	$250,000	$450,000	$0
BENEFITS	REDUCED UNEMPLOYMENT	$200,000	$50,000	$50,000	$100,000
	REDUCED ABSENTEEISM	500,000	0	400,000	100,000
	INCREASED PRODUCTIVITY	300,000	0	300,000	0
	TOTALS	$1,000,000	$50,000	$750,000	$200,000

FIGURE 4.7   GROUPINGS BY INTEREST (**Explanation**: Costs and benefits can also be separated and presented according to the principal actors or most relevant groups in the analysis. This example derives from the data shown in Figure 4.2.)

information placed under the headings *do nothing* or *new proposal* that do not yet warrant a complete analysis.

If, however, the *do nothing* option is a more serious matter, a full comparable benefit-cost analysis should be presented so that this option is considered as a real alternative (Figure 4.9). If the *new proposal* option is being given serious attention, a full comparable benefit-cost analysis is, by definition, impossible, as the proposal does not yet exist. What can be presented, however, is an estimated range of possibilities. The upper and lower estimates of benefits, costs, and effects for a *new proposal* should be designed to coincide with key values of the other proposed action(s). Under the title *new proposal*, the analyst should indicate at what level benefits must exceed costs to make this choice clearly more desirable. The presentor should also show the level of benefits and costs at which selecting the *new proposal* option is clearly less desirable.

CRITERIA FOR SELECTION

In most instances, a benefit-cost or cost effectiveness analysis entails a detailed comparison between at least two alternatives, even if one option is the *do-nothing* alternative. Criteria for choosing one action over another are not always self evident and may require their own supporting arguments if they are not to become controversial issues. This is especially true when there are several alternative actions with contingencies, uncertainties, and multiple payoffs. This section focuses on criteria used to choose from among relatively few, but complex, alternative actions, while the next chapter examines criteria for other types of decision-making structures and circumstances.

Ratios versus Absolute Differences

The conventional criteria for choosing an alternative is the benefit/cost ratio or effectiveness/cost ratio, where the benefits (or some measure of effectiveness) are divided by the costs. If the benefit/cost ratio is 2.5 for action *A* an 1.5 for action *B*, the former is selected. These ratios can be presented, and should be given clear visual emphasis in the table or chart portraying the analysis. The numbers usually are presented on the bottom or end of the table. It may be wiser, however, to present them on the top or left margin of the table, emphasizing them as the key decision criteria. This enables the audience to know where the argument will lead (Figure 4.10).

In most cases, the use of a benefit/cost or cost/effectiveness ratio is easy to justify and is commonly used when the costs of alternative actions are approximately equal. For example, an organization may have a predetermined or fixed budget that must be allocated. The fixed budget probably will represent the cost of any alternative. The benefits may vary, but the costs will

TABLE A. COSTS AND BENEFITS FOR PUBLIC HEALTH PROGRAM OPTIONS

	PROGRAM X HOME HEALTH CARE FOR THE ELDERLY	PROGRAM Y IMMUNIZATION PROGRAM FOR SCHOOL CHILDREN	OTHER OPTIONS
COSTS (IN MILLIONS OF DOLLARS)			
A. STAFF	2.6	4.1	If no new program is adopted, then the current programs for immunization and home health care will continue. Other options have also been explored but are not considered economically feasible.
B. FACILITY RENTAL/ FINANCIAL COSTS	0.8	0.5	
C. SUPPLIES AND EXPENSES	0.5	1.0	
D. OVERHEAD	1.0	1.3	
TOTAL ANNUAL COST	4.9	6.9	
BENEFITS (IN MILLIONS OF DOLLARS)			
A. SAVIINGS IN HOSPITAL STAFF AND TREATMENT COSTS	1.5	2.9	
B. SAVINGS IN OPERATIONAL COSTS FOR EXISTING FACILITIES	2.2	2.0	
TOTAL	3.7	4.9	

TABLE B: COSTS AND ORIGINAL NON-COMPARABLE BENEFITS

	OPTION A JOB CREATION & ECONOMIC DEVELOPMENT PROGRAM	OPTION B INDUSTRIAL HEALTH CARE PROGRAM	OTHER OPTIONS
COSTS: ANNUAL COST	$1,800,000	$700,000	If no new program option is selected then the current job retention program will continue. There is no currently operating industrial health care program. Other options for public investment have yet to be explored.
BENEFITS: ECONOMIC BENEFITS	A. 100 NEW JOBS/ YEAR B. GROWTH OF ECONOMIC BASE C. REDUCED UNEMPLOYMENT RATE	A. REDUCED ABSENTEEISM B. INCREASED WORKER PRODUCTIVITY	
SOCIAL BENEFITS	A. IMPROVEMENT IN POOR NEIGHBOR- HOODS B. IMPROVED BUSINESS MORALE	A. IMPROVED QUALITY OF LIFE B. IMPROVED JOB SATISFACTION	

FIGURE 4.8 REFERENCING OTHER OPTIONS (**Explanation**: In most analyses of the costs and benefits of alternative actions there are usually additional alternatives, such as "doing nothing" or developing an unexplored option. Sometimes the analyst may wish to reference these possibilities but not give them undue emphasis. Table A does this for the data shown in Figure 4.1 whereas Table B does this for Figure 4.2.)

		OPTION A: JOB CREATION AND ECONOMIC DEVELOPMENT PROGRAM	OPTION B: INDUSTRIAL HEALTH CARE PROGRAM	DO-NOTHING: CONTINUE EXISTING JOB PROGRAM	ALTERNATIVE INVESTMENTS (TAX REDUCTION INCUBATOR INDUSTRIES, ETC.)
COSTS	STAFF	$1,000,000	$150,000	$50,000	PRELIMINARY ESTIMATES SUGGEST OTHER OPTIONS COULD BE DEVELOPED WHICH WOULD RANGE IN COSTS FROM 500,000 TO 2,500,000
	FACILITIES	200,000	300,000	10,000	
	OVERHEAD	600,000	250,000	50,000	
	TOTALS	$1,800,000	$700,000	$110,000	
BENEFITS	NEW JOBS	$200,000	$0	$100,000	THE PRELMINARY ESTIMATES OF BENEFITS FROM THESE OTHER OPTIONS ARE ROUGHLY $700,000 TO $1,500,000
	REDUCED UNEMPLOYMENT	500,000	200,000	0	
	REDUCED ABSENTEEISM	0	500,000	0	
	INCREASED PRODUCTIVITY	0	300,000	0	
	TOTALS	$2,500,000	$1,000,000	$100,000	

FIGURE 4.9 ELABORATING *DO-NOTHING* AND *NEW PROPOSAL* OPTIONS (**Explanation**: In some situations the possibility of doing nothing or explaining some unknown option may deserve elaboration. This illustration uses data from Figures 4.2 and 4.7.)

be roughly equal. This is an appropriate situation for using a ratio as the choice criterion.

There are, however, other instances where the costs are not roughly equal. In these cases, the use of ratios as the most appropriate decision rule is debatable. If action *A* has a benefit/cost ratio of 2.5 ($250,000/$100,000) but action *B* has a benefit/cost ratio of 5.0 ($50,000/$10,000), action *B* would be selected if ratios are used. On the other hand, *the absolute difference* in the benefit minus the costs is $150,000 for action *A* but only $40,000 for action *B*. This criterion leads to the opposite conclusion and favors action *A* (Figure 4.11).

Using the absolute difference as a criterion for choosing actions is easy to explain and to comprehend. It does, however, create some new problems, one of which concerns the potential, but unexplored value of cost savings associated with one action relative to another. In the foregoing case, action *B* saves $90,000 more than action *A* ($150,000-$40,000), but if the $90,000 associated with action *B* versus *A* could be effectively used for some other program, those potential costs and benefits could be attributed to action *B*. This would then allow for a comparison of two choices, both with equal costs. However, the benefits of action *B* are now expanded, presenting a new analytic problem, as there are now roughly equal costs for each alternative set of actions. In fact, the analyst now can use a ratio criterion instead of the absolute difference in values (Figure 4.12).

TABLE A. RECOMMENDATION FOR HOME HEALTH CARE PROGRAM

	PROGRAM X HOME HEALTH CARE FOR THE ELDERLY	PROGRAM Y IMMUNIZATION PROGRAM FOR SCHOOL CHILDREN
BENEFIT/ COST RATIO	.76	(.71)
COSTS (IN MILLIONS OF DOLLARS)		
A. STAFF	2.6	4.1
B. FACILITY RENTAL/ FINANCIAL COSTS	0.8	0.5
C. SUPPLIES AND EXPENSES	0.5	1.0
D. OVERHEAD	1.0	1.3
TOTAL ANNUAL COST	4.9	6.9
BENEFITS (IN MILLIONS OF DOLLARS)		
A. SAVIINGS IN HOSPITAL STAFF AND TREATMENT COSTS	1.5	2.9
B. SAVINGS IN OPERATIONAL COSTS FOR EXISTING FACILITIES	2.2	2.0
TOTAL	3.7	4.9

TABLE B. RECOMMENDATION FOR HOME HEALTH CARE PROGRAM

		OPTION A: JOB CREATION AND ECONOMIC DEVELOPMENT PROGRAM	OPTION B: INDUSTRIAL HEALTH CARE PROGRAM	DO-NOTHING: CONTINUE EXISTING JOB PROGRAM	ALTERNATIVE INVESTMENTS: (TAX REDUCTION INCUBATOR INDUSTRIES, ETC.)
BENEFIT/ COST RATIO		1.39	(1.43)	0.91	ESTIMATED RANGE 1.40 TO 0.70
COSTS	STAFF	$1,000,000	$150,000	$50,000	PRELIMINARY ESTIMATES SUGGEST OTHER OPTIONS COULD BE DEVELOPED WHICH WOULD RANGE IN COSTS FROM 500,000 TO 2,500,000
	FACILITIES	200,000	300,000	10,000	
	OVERHEAD	600,000	250,000	50,000	
	TOTALS	$1,800,000	$700,000	$110,000	
BENEFITS	NEW JOBS	$200,000	$0	$100,000	THE PRELMINARY ESTIMATES OF BENEFITS FROM THESE OTHER OPTIONS ARE ROUGHLY $700,000 TO $1,500,000
	REDUCED UNEMPLOYMENT	500,000	200,000	0	
	REDUCED ABSENTEEISM	0	500,000	0	
	INCREASED PRODUCTIVITY	0	300,000	0	
	TOTALS	$2,500,000	$1,000,000	$100,000	

FIGURE 4.10 BENEFIT/COST RATIOS (*Explanation*: The more conventional criterion used to recommend a decision is the benefit/cost ratio (it is calculated simply by dividing total benefits by total costs). These data should be featured prominently in the presentation. Table A does this for Figure 4.1A. Table B does this for Figure 4.9.)

In this example, action *A* remains intact. Action *B*, however, has changed. Its cost has increased to match the cost of action *A* and, more importantly, its benefits are expanded to include new types of outcomes. In effect, action *B* is a compound action with two implied sets of costs and benefits, and the total cost is set equal to a budget limitation. Even more complicated situations arise when there are several actions that can be combined in different ways within

TABLE A. RECOMMENDATION FOR JOB CREATION/ ECONOMIC DEVELOPMENT PROGRAM

NOTE: GIVEN THE RELATIVE PROXIMITY OF THE BENEFIT/ COST RATIOS,
BUT THE SUBSTANCIAL DIFFERENCE IN THE SIZE OF THE NET
BENEFITS BETWEEN THE TWO OPTIONS, OPTION A IS RECOMMENDED

		OPTION A: JOB CREATION AND ECONOMIC DEVELOPMENT PROGRAM	OPTION B: INDUSTRIAL HEALTH CARE PROGRAM	DO-NOTHING: CONTINUE EXISTING JOB PROGRAM	ALTERNATIVE INVESTMENTS: (TAX REDUCTION INCUBATOR INDUSTRIES, ETC.)
BENEFIT- COSTS		$700,000	$300,000	-$50,000	ESTIMATED RANGE +$200,000 TO -$1,000,000
BENEFIT/ COST RATIO		1.39	.1.43	0.91	ESTIMATED RANGE 1.40 TO 0.70
COSTS	STAFF	$1,000,000	$150,000	$50,000	PRELIMINARY ESTIMATES SUGGEST OTHER OPTIONS COULD BE DEVELOPED WHICH WOULD RANGE IN COSTS FROM 500,000 TO 2,500,000
	FACILITIES	200,000	300,000	10,000	
	OVERHEAD	600,000	250,000	50,000	
	TOTALS	$1,800,000	$700,000	$110,000	
BENEFITS	NEW JOBS	$200,000	$0	$100,000	THE PRELMINARY ESTIMATES OF BENEFITS FROM THESE OTHER OPTIONS ARE ROUGHLY $700,000 TO $1,500,000
	REDUCED UNEMPLOYMENT	500,000	200,000	0	
	REDUCED ABSENTEEISM	0	500,000	0	
	INCREASED PRODUCTIVITY	0	300,000	0	
	TOTALS	$2,500,000	$1,000,000	$100,000	

FIGURE 4.11   PRESENTING ALTERNATIVE DECISION-MAKING CRITERIA AND RECOMMENDA-TIONS   (**Explanation**: In cases where the costs are not roughly equal, it may be unwise to use the benefit/cost ratio as the principal or only decision-making criterion. The actual difference (benefits minus costs) may be a reasonable option in this situation. Table A presents these two criteria for data shown in Figure 4.10, along with an argument that favors rejecting the ratio criterion. In the example in Table B, an argument is created for accepting the ratio criterion as described in the text.)

a given budget. This is an entirely different problem concerning the allocation of resources, and has its own logic, computational techniques, and shortcomings.

Some issues regarding allocation of resources are discussed in Chapter Six in relation to the distribution of time, money, and selected resources for prompt management. However, formal computational algorithms for allocation of some resources to achieve measurable objectives are especially complex, difficult to present, and infrequently used relative to other available techniques that have been discussed.

TABLE B.   RECOMMENDATION FOR INVESTMENT A

	INVESTMENT A ADD NEW PUBLIC FACILITY WITH BROADER MARKET	INVESTMENT B MODIFY EXISTING FACILITY TO ALLOW FOR TICKET BOOTH
BENEFIT/ COST RATIO	2.5	5.0
BENEFIT/ COSTS	150,000	40,000
BENEFITS NUMBERS OF USERS	50,000	10,000
REVENUE	250,000	50,000
TOTALS	250,000	50,000
COSTS ANNUAL OPERATION OF FACILITY	20,000	10,000
AMORTIZATION OF CAPITAL	80,000	0
TOTALS	100,000	10,000

NOTE: ALTHOUGH THE ACTUAL NET BENEFIT FOR INVESTMENT A IS SUBSTANTIALLY HIGHER, IT REQUIRES AN EQUALLY SUBSTANTIAL INVESTMENT. INVESTMENT B PROVIDES BOTH A BETTER RATE OF RETURN AND THE POSSIBILITY OF INVESTING $90,000 IN OTHER ENDEAVORS (PRESUMING THAT OTHER ATTRACTIVE OPTIONS CAN BE FOUND).

Ratios with Noncomparable Values

Thus far, the examples in the text have focused only on criteria where benefits and costs are measured in comparable terms, specifically money, which is easy for audiences to comprehend. In some cases, such as cost-effectiveness ratios, the presentation becomes more ambiguous, and the presentor is often dividing apples by oranges. For example, the ratio may be a *unit service cost* such as the number of persons or *units* treated in a health program divided by the *cost* of the program. *Student/teacher* ratios present a similar situation. The number of students instructed (the effect) is divided by the number of faculty required (the cost). Some ratios are reasonably self evident to the audience (Figure 4.13), but others may require special explanation or illustration, especially when the ratios are based on large, indivisible entities.

In a situation where two alternative programs exist for locating new sales offices (or government information centers or service bureaus), the effects and costs may be measured in large, indivisible entities, such as the number of communities serviced (effects), and the number of sales offices required to provide service (costs). The effectiveness/cost ratio would be *communities per office*. If there is an effectiveness/cost ratio of 1.8 for program *A* and 1.6 for program *B*, it may prove to be too abstract for the audience. Program *A*, with the higher ratio, may include just 1 office that serves 20 communities and 14 other offices that each serve only 1 community. Program *B* may have 25 offices that each serve either 1, 2, or 3 communities. In other words, it is not just the ratios that are different, but the entire structure of the two programs (Figure 4.14).

The point is that with large, indivisible effects and costs, there are hidden factors that raise questions about the significance and validity of comparing ratios. As soon as an audience sees an indivisible phenomenon presented mathematically as if it were divisible, there is likely to be some doubt and speculation. Even the ubiquitous cliche of *2.5 persons per family* makes some audiences want to know how many are 2-person families, 3-person families, and so on.

One way to address the situation is to show the basis for the statistics and potential ambiguities. In the forementioned case, the analyst could present, adjacent to the principal table, a statement concerning the range and distribution of communities served by each local sales office. At the very least, there should be examples of what the ratios mean in concrete terms, such as a small table that shows which office corresponds to each community.

Indivisibilities are an invitation to arguments against the validity of the analysis. These arguments usually can be addressed satisfactorily, but require special attention in the presentation. This does not mean the ratio criterion should be avoided, as it may still be valid. Rather, full explanations and justifications may be needed to convince the audience of the validity of the figures presented.

TABLE A.

	INVESTMENT A (TOTAL)	INVESTMENT B (TOTAL)	INDIVIDUAL COMPONENTS IN INVESTMENT B		
	ADD NEW PUBLIC FACILITY WITH BROADER MARKET	FACILITY MODIFICATIONS AND DEBT REDUCTION	MODIFY EXISTING FACILITY TO ALLOW FOR TICKET BOOTH	MODIFY SECOND FACILITY TO ALLOW FOR ADMISSION CHARGE	INVEST IN PUBLIC DEBT REDUCTION PROGRAM
BENEFIT/ COST RATIO	(2.5)	2.3	NOT APPLICABLE	NOT APPLICABLE	NOT APPLICABLE
BENEFIT/ COSTS	150,000	130,000			
BENEFITS					
REVENUE	250,000	90,000	50,000	40,000	0
OTHER	0	140,000	0	0	140,000
TOTALS	250,000	230,000	50,000	40,000	140,000
COSTS					
ANNUAL OPERATION	20,000	25,000	10,000	10,000	5,000
AMORTIZATION	80,000	5,000	0	5,000	0
OTHER	0	70,000	0	5,000	65,000
TOTALS	100,000	100,000	10,000	20,000	70,000

TABLE B

	INVESTMENT A	INVESTMENT B
	ADD NEW PUBLIC FACILITY WITH BROAD MARKET APPEAL	CONSTRUCT TICKET BOOTH AND INVEST REMAINDER IN INTEREST BEARING ACCOUNT
BENEFIT/ COST RATIO	2.5	(2.7)
BENEFITS (NET REVENUE)	250,000	270,000=(50,000 + 220,000)
COSTS	100,000	100,000=(10,000 + 90,000)

FIGURE 4.12 CREATING COMPARABLE BENEFIT/COST RATIOS (**Explanation**: As a decision-making criterion the benefit/cost ratio is theoretically superior when each option has roughly equal costs; that is, the denominators in the ratios are equivalent. Tables A and B show two ways to do this for the data in Figure 4.11. In Table A this results in a recommendation for investment whereas in Table B the resulting recommendation is a new *compound* investment B.)

TABLE A.   RECOMMENDATION FOR IMMUNIZATION PROGRAM BASED ON EFFECTIVENESS RATIOS

	PROGRAM X HOME HEALTH CARE FOR THE ELDERLY	PROGRAM Y IMMUNIZATION PROGRAM FOR SCHOOL CHILDREN
EFFECTIVENESS/ COST RATIO (ADDED LIFE YEARS/ COST IN $1,000)	(2.61)	.61
EFFECTIVENESS ADDED LIFE YEARS	18,000 (0.1 x 180,000)	3,000 (0.2 x 15,000)
AVERAGE INCREASE IN LIFE EXPECTANCY	0.1	0.2
COSTS (IN $1,000)	$6,900	$4,900

TABLE B.   RECOMMENDATION FOR PUBLIC FACILITY ADDITION

	INVESTMENT A ADD NEW PUBLIC FACILITY WITH BROADER MARKET	INVESTMENT B MODIFY EXISTING FACILITY TO ALLOW FOR TICKET BOOTH
EFFECTIVENESS/ COST RATIO (ADDED LIFE YEARS/ COST IN $1,000)	0.50	1.00
EFFECTIVENESS TOTAL NUMBER OF USERS	50,000	10,000
COSTS ANNUAL OPERATION OF FACILITY	20,000	10,000
AMORTIZATION OF CAPITAL	80,000	0
TOTALS	100,000	10,000

FIGURE 4.13 EFFECTIVENESS/ COST RATIOS (**Explanation**: Sometimes the decision-making criterion relates to qualitative issues instead of, or in addition to, just dollar comparisons as expressed in a benefit/cost ratio. Table A shows how this concept might be applied to the data in Figure 4.1. Table B does this for the data in Figure 4.11.)

TABLE A. RECOMMENDATION FOR SALES OFFICE DISTRIBUTION

	PROGRAM A	PROGRAM B	PROGRAM C
EFFECTIVENESS/ COST RATIO (communities served per office)	1.8	1.6	1.5
NUMBER OF COMMUNITIES THAT CAN BE SERVED SATISFACTORILY	27	40	30
NUMBER OF OFFICES CONSTRUCTED (AVE. ANNUAL COST OF $100,000)	15	25	20

TABLE B. ANALYSIS OF SALES OFFICE DISTRIBUTION

	PROGRAM A		PROGRAM B		PROGRAM C	
	URBAN MARKET AREAS	SUB-URBAN MARKET AREAS	URBAN MARKET AREAS	SUB-URBAN MARKET AREAS	URBAN MARKET AREAS	SUB-URBAN MARKET AREAS
EFFECTIVENESS/ COST RATIO (communities served per office)	2.0	1.8	2.5	1.0	2.0	1.33
NUMBER OF COMMUNITIES THAT CAN BE SERVED SATISFACTORILY	2	25	25	15	10	20
NUMBER OF OFFICES CONSTRUCTED (AVE. ANNUAL COST OF $100,000)	1	14	10	15	5	15

FIGURE 4.14 USING EFFECTIVNESS/COST RATIOS WITH NONMONETARY COSTS (**Explanation**: In some cases the costs and the effects of alternative options are quantified in nonmonetary terms. In these two tables the same data are presented, but in Table B it is a disaggregated to create a more complex image, which may have a different impact on the audience.)

5

Statistics For Decisions:
Matrices And Hierarchies

Almost all statistical presentations relate to making decisions. Inferential statistics—that is, statistics concerned with testing hypotheses—help audiences make decisions regarding the truth or falsity of a statement. Other statistical methods focus directly on the pragmatic rather than the truth value of alternative actions. This latter approach to statistics is found in business, planning, engineering, and various policy sciences.

The distinction between deciding the truth or falsity of a hypothesis as opposed to deciding on a pragmatic course of action is critical. For example, there are different branches or philosophies of statistics underlying these two types of decisions. *Non-Bayesian* statistics form the basis for some inferential statistics and hypothesis testing, whereas *Bayesian* statistics are used in the approach often found in business decision making, operations research, planning, and related disciplines.

The difference between the two approaches is based on a controversial theorem called Bayes' theorem. This allows for the inclusion of subjective judgment in the computation of probabilities used in making decisions. The premise in inferential statistics is that such subjective judgments about the *prior* probabilities of events are totally unwarranted. The pragmatic approach maintains that prior knowledge about a subject, especially practical experience, should be used in making a decision, and to do otherwise would increase the risk of making poor and/or unprofitable decisions.

Resolving such philosophical differences is not an intent of this book. However, it must be noted that inferential statistics very often are misused to justify a pragmatic course of action rather than being limited to establishing the validity of a hypothesis. For instance, if an audience is shown an analysis supporting the inference that income levels are related to education, the analysis should support only the hypothesis. It does not, say, support a course of action to improve income levels by improving educational programs. Any course of action should not be presumed appropriate solely because it is associated with valid statements derived from proper inferential statistics.

The issue for the presentors of statistics is to maintain their credibility and their responsibility to the audience by clearly separating decisions regarding the merits of an action that may be based, in part, on statistical analysis, from decisions concerning the validity of an inference. Techniques for presenting inferences, along with other descriptive statistics, were discussed in Chapters One, Two, and Three, and are appropriate, useful components of arguments leading to practical decisions but are not, in themselves, decision-making arguments.

This chapter is concerned with statistical decision-making arguments and effective techniques for their presentation. Chapter Four alluded to some of this material, although the methods discussed there were linked to a different conceptual base. Discussions of costs, benefits, and effects focused more on the problem of measuring a few complex alternative actions, and the decision criteria were comparatively simple, little attention being given to the uncertainty and probability of the outcomes.

Formal theories for statistical decision making have a different focus, and more actions have to be considered. Application of decision criteria is more complex, and uncertainties and probabilities play a crucial role. The presentation of statistical decision-making methods has traditionally taken several forms, the most useful of which are the decision matrix and the decision tree, or hierarchy. These two presentation formats form the basis for the next two sections, which are followed by a separate section on *Ranking and Weighting*.

THE DECISION MATRIX

Historically, statistical decision theory grew out of business theory and operations research. In these disciplines, specific statistical methods are used to compare the value of different actions, to account for the probabilities of obtaining those values and to apply criteria for choosing the most appropriate course of action. A number of texts elaborate on these theories and procedures (see Appendix for references), but the purpose of this chapter is to describe how such statistical methods, explanatory statements, and arguments can be effectively presented to a general audience. Consequently, this and subsequent sections are organized according to presentation issues rather than the sequence of computational formulas and principles on which statistical decision theory is based.

The Basic Decision Matrix: Actions, Events, Probabilities, Comparable Payoffs

One of the most common presentation techniques is a decision matrix, which is analogous to the use of two-way tables for descriptive statistics. It is easy to prepare and, more importantly, easy to comprehend. Typically, one side of the table lists alternative actions from which the presentor will recommend one action (or a subset of actions). The other dimension of the table or matrix may vary, but often shows alternative events or states of nature. The individual cells in the matrix display outcomes or payoffs. The outcomes have numeric values, which often are expressed in monetary terms, where each outcome corresponds to the action and event from which it derives (Figure 5.1). A decision matrix can take many forms and contents, but in its simplest computational form would have the following attributes:

1. A set of several discrete actions (usually five to ten)
2. A set of several events which may occur and which are mutually exclusive and collectively exhaustive; that is, if one event occurs none of the other events can occur, *and* all possible events are included
3. Probabilities for each event; that is, the likelihood that an event will occur (the probabilities add up to 1.0)
4. A set of discrete values, typically monetary, that will be obtained for each action if each event occurs

This kind of matrix is most useful when all the data are available, and all the logical constraints can be satisfied. In practice, this is rarely true, since missing data and ambiguous situations are likely to complicate the issue. Nevertheless, this form of matrix is still widely used because it is conceptually simple and easy to comprehend. It is also a good starting point for this discussion, because it illustrates many basic presentation difficulties.

Initially there should be clear labelling of all the actions, events and payoffs. Simply using codes (such as Action *A*, Action *B*, ...Event *A*, Event *B*, ...) may be too abstract for most audiences. Similarly, each of the payoffs or outcomes should be explained, and if there are too many potential outcomes, there should at least be some sample derivations.

A problem may arise in explaining probabilities to a general audience. Most audiences have not studied statistical decision theory and may feel uncomfortable with the image of gambling that emerges when probabilities are used. Consequently, the presentation must display the derivation of the probabilities, even if they are based exclusively on expert judgments.

TABLE A   ABSTRACT MATRIX

EVENTS
(and Probabilities)

	E1	E2
Probability ➤	0 .50	0 .50
A1	1.0	-0.8
A2	-0.2	+0.6

ACTIONS (left side label)

Sum of Probabilities= 1.0

Payoffs (in millions of dollars)

EVENTS (E1 and E2): Mutually exclusive and collectively exhaustive with probabilities that sum to 1.0.

ACTIONS (A1 and A2): The alternative decisions, one of which must be selected.

PAYOFFS: Each cell shows the outcome, (in this case, financial payoffs) which wil occur if the corresponding decision (row) is selected and the corresponding event (column) occurs.

TABLE B   APPLIED EXAMPLE

HEALTH PLAN ENROLLMENT PREDICITONS

	Low Enrollment	Moderate Enrollment	High Enrollment
Probability ➤	0.20	0.50	0.30
No New Construction	0.5	1.0	- 0.5
Single New Major Facility	- 4.0	- 2.0	14.0
New Satellite Facilities	- 3.0	6.0	12.0
Staged Satellite Development	- 0.6	3.2	7.5

CLINIC DEVELOPMENT OPTIONS (left side label)

(Payoffs shown in $millions/ year in annual revenue)

FIGURE 5.1 BASIC DECISION MATRIX (*Explanation:* Table A is an abstract example of a basic decision matrix containing sample calculations. Table B is an applied example.)

More problematic is the explanation of the logical interrelation between probabilities and events. In statistical decision theory, the *expected* value of an action depends upon the product of the payoff times the probability. Where a program has the potential or unconditional value of generating $1,000,000 in revenue if a particular event occurs, it has an expected value of only $500,000 if the event has only a 50 percent chance of occurring ($1,000,000 × .50 = $500,000). More precisely, the expected value of an action is the sum of all potential values multiplied by the respective probabilities that they will actually occur. Thus, a program that has a 50 percent chance of generating $1,000,000 in revenue and a 50 percent chance of losing $800,000 in revenue actually has a total expected value of $100,000 (Figure 5.2).

An audience viewing such an analysis might feel uncomfortable if the presentor simple asserted that taking this particular action was worth $100,000. In fact, it has either a gain of $1,000,000 *or* a loss of $800,000. Consequently, any analysis that presents the sum of probabilities times values has to include some detailed, convincing examples of the legitimacy of the procedure.

Using the sum of expected values is only one simplified approach for choosing an action, and may not fit the issue and/or the audience. There are several other procedures for selecting among alternative actions that are considered in the following subsections.

Decision Criteria

Given a matrix with actions, events, probabilities, and measurable payoffs, there are many rules or algorithms for selecting the wisest course of action. These rules usually are called *decision criteria*, and their proper presentation is critical. The same matrix of data can lead to entirely different recommendations, depending on the decision criterion being applied.

One such criterion, the sum of expected values, has been noted in the previous subsection. In that hypothetical example, there were large positive and large negative payoffs for a particular action. In practice, avoiding a large negative payoff may be a more appropriate decision criterion than simply selecting the action with the largest total expected value.

This latter approach is referred to as *maximizing* the expected value. Alternatively, avoiding the worst possible outcome is sometimes referred to as *minimizing the maximum loss*. That is, the method requires noting the worst thing that could happen under any circumstances for each action, and then picking the action that has the so-called best of the possible bad outcomes (Figure 5.3).

Taking this line of reasoning one step further, there is a criterion referred to as *minimizing the expected maximum loss*. In this case, the actual payoffs are not used, and the decision is based on the lowest products of each payoff multiplied by the probability that it might occur. The rationale for using this approach is complex. For example, two programs might, under the most

TABLE A   ABSTRACT MATRIX

		EVENTS		
		E1	E2	TOTAL EXPECTED VALUE
	Probability →	0.50	0.50	
A C T I O N S	A1 Not Recommended	$1,000,000 x 0.5= $500,000	$-800,000 x 0.5= $-400,000	+ 500,000 - 400,000 $ 100,000
	A2 Recommended: Highest Total Expected Value	$- 200,000 x 0.5= $- 100,000	$ 600,000 x 0.5= $300,000	- 100,000 + 300,000 $ 200,000

TABLE B   RECOMMENDATION FOR CLINIC DEVELOPMENT
                 (BASED ON EXPECTED VALUE)

		HEALTH PLAN ENROLLMENTS			
		Low Enrollment	Moderate Enrollment	High Enrollment	TOTAL EXPECTED VALUE
	Probability →	0.20	0.50	0.30	
No New Construction	Base Estimate	500,000	1,000,000	-500,000	
	Expected Value	$100,000	$500,000	$-150,000	$450,000
Single New Major Facility	Base Estimate	- 4,000,000	-2,000,000	14,000,000	
	Expected Value	$-800,000	$-1,000,000	$4,200,000	$2,400,000
New Satellite Facilities	Base Estimate	-3,000,000	6,000,000	12,000,000	
	Expected Value	$-600,000	$3,000,000	$3,600,000	$6,000,000
Staged Satellite Development	Base Estimate	- 600,000	3,200,000	7,500,000	
	Expected Value	$-120,000	$1,600,000	$2,250,000	$3,730,000

FIGURE 5.2   PRESENTING EXPECTED VALUES   (*Explanation:* The elements of a traditional decision matrix are often used to compute the *expected value* of each action, which is then used as a basis for recommending the decision with the highest total expected value. Table A illustrates this, showing the computational procedures. Table B is an applied example. Both derive from Figure 5.1.)

adverse circumstances, produce losses of $1,000,000 (program *A*) or $200,000 (program *B*). The probability of such adverse circumstances may, however, be only 1 percent for program *A* and 10 percent for program *B*. Thus, the expected maximum loss is $10,000 for program *A* and $20,000 for program *B*. The criterion of minimizing maximum expected loss would lead to selecting *A*, whereas just minimizing maximum loss would lead to selecting *B* (Figure 5.3). This is a simple issue for an analyst trained in statistical decision theory, but can

TABLE A   ABSTRACT EXAMPLE

ACTIONS			EVENTS				Worst Possibility
			E1	E2	E3	E4	
	Probability:		.01	.10	.25	.64	
	Action A	Actual Value	- 1,000,000	- 50,000	0	+ 800,000	- 1,000,000
		Expected Value	*(- 10,000)*	(-5,000)	(0)	(+ 512,000)	*(- 10,000)*
	Action B	Actual Value	0	*-200,000*	-100,000	+ 700,000	*-200,000*
		Expected Value	(0)	(- 20,000)	(- 25,000)	(+ 448,000)	(- 25,000)

Action 'A' has the smaller (minimum) <u>expected</u> loss

Action 'B' has the smaller (minimum) possible (or unconditional) loss

Note: The worst actual value for action A is -1,000,000, and for Action B the worst actual value is -200,000. Therefore, to minimize the maximum possible loss, action B should be selected. The <u>expected losses</u> are different: for action A the worst expected loss (-10,000) is lower than the worst expected loss for action B (-20,000).

TABLE B   APPLIED EXAMPLE

		HEALTH PLAN ENROLLMENTS			
		Low Enrollment	Moderate Enrollment	High Enrollment	Worst Outcome for Each Action
	Probability	0.20	0.50	0.30	
No New Construction	Actual Value	500,000	1,000,000	-500,000	-500,000
	Expected Value	$100,000	$500,000	$-150,000	$-150,000
Single New Major Facility	Actual Value	- 4,000,000	-2,000,000	14,000,000	- 4,000,000
	Expected Value	$-800,000	$-1,000,000	$4,200,000	$- 1,000,000
New Satellite Facilities	Actual Value	-3,000,000	6,000,000	12,000,000	- 3,000,000
	Expected Value	$-600,000	$3,000,000	$3,600,000	$-600,000
Staged Satellite Development	Actual Value	- 600,000	3,200,000	7,500,000	-600,000
	Expected Value	$-120,000	$1,600,000	$2,250,000	$- 120,000

    Action with the best of the <u>worst possible outcomes</u>
    Action with the best of the <u>worst expected outcomes</u>

FIGURE 5.3 AVOIDING THE WORST OUTCOMES (***Explanation:*** A common criterion for making some decisions is avoiding the worst outcomes. It is based on the pessimistic attitude that presumes for each action the worst outcome (maximum loss) will occur and, therefore, the decision maker should select the action with the minimum loss. Minimizing the maximum loss is sometimes called a *minimax* criterion. It can also be applied to the expected values of each outcome as well as to the absolute value. Table A contains an abstract example, and Table B shows an applied example.)

be very confusing for a general audience.

Presenting these legitimate approaches to selecting actions is quite difficult, as they involve complex computations and, more importantly, complex rationales. Minimizing potential losses, for instance, regardless of the approach, is essentially a pessimistic rationale. It rests largely upon examination of worst cases. There are, however, converse rationales and decision criteria that are essentially optimistic, such as *maximizing the maximum gain*, which involves picking the action with the best possible payoff regardless of the potential for lower or negative payoffs. There is also a criterion of maximizing the expected maximum gain, which again involves the use of probabilities.

The rationality of these approaches is debatable, but generally the use of decision matrices can be helpful if there is a clear presentation of the technique, and if special emphasis is given to the decision criterion. In some cases, it may be appropriate to offer more than one decision criterion to the audience, especially if the presentor is only recommending actions for audience consideration. Consequently, it may be appropriate to show that an "expected value" criterion leads to one choice while *minimizing maximum loss* leads to a different choice. This allows the audience to discuss which criterion best suits their needs (Figure 5.4).

Ambiguous Events and Probabilities

There are many situations in which the events, probabilities, and data do not fit the logical constraints and computational procedures of a decision matrix; for example, some of the events may be unknown, they may not be mutually exclusive, or the probabilities cannot be accurately measured.

The analyst always has the choice of forcing the situation to fit the matrix, thus gaining the advantage of having a potentially neat and tidy decision matrix. There is, however, a simultaneous risk of presenting information that is less relevant and potentially misleading; hence, forcing the information to fit an analytic model may be appropriate only if there are minor distortions. The alternative approach is to abandon the formal constraints of the analytic procedure to increase the relevance and appropriateness of the information being presented. Some common problems are discussed in the following subcategories.

Unknown Probabilities
A typical problem occurs when precise probabilities are unknown. In this situation, it may be effective to present the probabilities of events as *high, moderate,* or *low,* or to group events into such categories as *highly likely events* and *unlikely events* (Figure 5.5).

Another approach would utilize a numerical range for the probability of an event. For example, event X may have a probability of 0.2 to 0.4, event Y a probability of 0.4 to 0.8, and so on. Use of such ordinal categories or ranges creates major logical problems, because the sum of the probabilities no longer equals 1.0. This is a requirement in some methods, especially since the sum of the probabilities for mutually exclusive and collectively exhaustive events must, by definition, equal 1.0. Consequently, when one is using ranges of probabilities, it may be appropriate to present two sets of estimates, each of which totals 1.0. By using this approach, it may be possible to apply more rigorous decision algorithms, although there will be two (or more) sets of

TABLE A   DECISION-MAKING OPTIONS FOR CLINIC DEVELOPMENT

HEALTH PLAN ENROLLMENTS

ACTION	RATIONALE	LOW ENROLLMENT	MODERATE ENROLLMENT	HIGH ENROLLMENT	
		0.20	0.50	0.30	◄─Probability
NO NEW CONSTRUCTION	This action minimizes the possible loss : the worst outcome is a loss of $500,000 (all other actions have higher possible losses)	500,000	1,000,000	-500,000	Actual Value
		$100,000	$500,000	$-150,000	Expected Value
SINGLE NEW MAJOR FACILITY	This action has the highest possible gain of $14,000,000 and the highest expected gain of $4,200,000 (no other actions have such high gains)	- 4,000,000	-2,000,000	14,000,000	Actual Value
		$-800,000	$-1,000,000	$4,200,000	Expected Value
NEW SATELLITE FACILITIES	This action has the highest total expected value of $6,000,000 (all other actions have lower total expected values)	-3,000,000	6,000,000	12,000,000	Actual Value
		$-600,000	$3,000,000	$3,600,000	Expected Value
STAGED SATELLITE DEVELOPMENT	This action minimizes the expected loss: the worst outcome is an expected loss of $-120,000 (all other actions have higher expected losses)	- 600,000	3,200,000	7,500,000	Actual Value
		$-120,000	$1,600,000	$2,250,000	Expected Value

FIGURE 5.4   PRESENTING DECISION CRITERIA   (*Explanation:* The analysis of the values associated with each action and event may be less important than the philosophy behind the decision-making criteria or rules for choice. The selection of one criterion over another is often subjective, and the analyst should therefore show different options. These two tables show how different sets of criteria may be represented. Table A illustrates a presentation that does not favor any one criterion. Table B shows how one criterion could be emphasized.)

expected outcomes corresponding to the two (or more) sets of probability estimates.

Events That Are Not Mutually Exclusive
Other ambiguities occur when events are not mutually exclusive. For example, the decision concerning an environmental health program may be contingent upon social events and environmental events that are not mutually exclusive.

TABLE B   RECOMMENDED ACTION FOR CLINIC DEVELOPMENT

HEALTH PLAN ENROLLMENTS

ACTION	RATIONALE	LOW ENROLLMENT	MODERATE ENROLLMENT	HIGH ENROLLMENT	←Probability
		0.20	0.50	0.30	
STAGED SATELLITE DEVELOPMENT (A)	This action is recommended. It minimizes the expected loss, it has only a slightly worse possible outcome than action B, and still offers reasonable gains	- 600,000	3,200,000	7,500,000	Actual Value
		$-120,000	$1,600,000	$2,250,000	Expected Value
NO NEW CONSTRUCTION (B)	While this action minimizes the possible loss, it is not substantially better than action A and offers lower, actual and expected gains.	500,000	1,000,000	-500,000	Actual Value
		$100,000	$500,000	$-150,000	Expected Value
SINGLE NEW MAJOR FACILITY (C)	While this action offers the greatest gain, the potential losses are too risky	- 4,000,000	-2,000,000	14,000,000	Actual Value
		$-800,000	$-1,000,000	$4,200,000	Expected Value
NEW SATELLITE FACILITIES (D)	While this action offers the highest total expected value, the potential loss is too risky	-3,000,000	6,000,000	12,000,000	Actual Value
		$-600,000	$3,000,000	$3,600,000	Expected Value

TABLE A   RECOMMENDED ACTION FOR CLINIC DEVELOPMENT

HEALTH PLAN ENROLLMENTS

ACTION	RATIONALE	MODERATE ENROLLMENT	HIGH ENROLLMENT	LOW ENROLLMENT
		RANGE OF PROBABILITIES		
		MOST PROBABLE	MID-RANGE PROBABILITY	LEAST PROBABLE
STAGED SATELLITE DEVELOPMENT (A)	This action is recommended. Given the rough probabilities it will have a low expected loss and, only a sightly worse possible outcome than action B, and still offers reasonable gains.	$3,200,000	$7,500,000	$- 600,000
NO NEW CONSTRUCTION (B)	While this action minizes the possible loss, it is not substancially better than action A and offers lower, actual and expected gains.	$1,000,000	$-500,000	$500,000
SINGLE NEW MAJOR FACILITY (C)	While this action offers the greatest gain, the potential losses are too risky	$-2,000,000	$14,000,000	$- 4,000,000
NEW SATELLITE FACILITIES (D)	While this action has a very high expected value, given the possible loss is too great compared to (A)	$6,000,000	$12,000,000	$-3,000,000

FIGURE 5.5 PRESENTING DECISIONS WITH UNKNOWN PROBABILITIES (**Explanation:** When precise probabilities are unknown, then the decision criterion can be applied only in a more subjective manner. Tables A and B show two ways to do this for the decision presented in Table B of Figure 5.4.)

There may be an increase in population (social event) with or without an improvement in air or water pollution (environmental event). One approach to this problem is to create compound events that are mutually exclusive (Figure 5.6). This can, however, become visually cumbersome and confusing if each compound event has three or more components. To avoid this, the decision matrix may be avoided, and a decision hierarchy that allows for contingent or conditional events may be used instead (this is discussed in the next section).

Of course, the analyst may want to retain the decision matrix for other reasons, such as creating an effective, understandable presentation. Other strategies are therefore needed to present decisions where the events are not

TABLE B   RECOMMENDED ACTION FOR CLINIC DEVELOPMENT

HEALTH PLAN ENROLLMENTS

ACTION	RATIONALE		MODERATE ENROLLMENT	HIGH ENROLLMENT	LOW ENROLLMENT
			RANGE OF PROBABILITIES		
			Moderate (0.4 to 0.7)	High (0.2 to 0.4)	Low (0.1 to 0.3)
STAGED SATELLITE DEVELOPMENT (A)	This action is recommended. Given the rough probabilities it will have a low expected loss and, only a sightly worse possible outcome than action B, and still offers reasonable gains.	Actual Value	$3,200,000	$7,500,000	$- 600,000
		Expected Value	1,280,000 to 2,240,000	1,500,000 to 3,000,000	-60,000 to -180,000
NO NEW CONSTRUCTION (B)	While this action minimizes the possible loss, it is not substancially better than action A and offers lower, actual and expected gains.	Actual Value	$1,000,000	$-500,000	$500,000
		Expected Value	400,000 to 700,000	-100,000 to -200,000	50,000 to 150,000
SINGLE NEW MAJOR FACILITY (C)	While this action offers the greatest gain, the potential losses are too risky	Actual Value	$-2,000,000	$14,000,000	$- 4,000,000
		Expected Value	-800,000 to -1,400,000	2,800,000 to 5,600,000	-400,000 to -1,200,000
NEW SATELLITE FACILITIES (D)	While this action has a very high expected value, given the rough probabilities, the potential loss is too great.	Actual Value	$6,000,000	$12,000,000	$-3,000,000
		Expected Value	2,400,000 to 4,200,000	2,400,000 to 4,800,000	-300,000 to -900,000

TABLE A   OUTCOMES FOR ALTERNATIVE HEALTH PROGRAMS

	Increase in Pollution <u>and</u> Increase in Population	Increase in Pollution <u>and</u> Population remains stable	No Increase in Pollution <u>and</u> Increase in Population	No Increase in Pollution <u>and</u> Population Remains Stable
PROBABILITY→	0.25 Moderate	0.35 High	0.10 Low	0.35 High
OPTION 1: Industrial Health Program	5500	5000	4800	4000
OPTION 2: Adult Screening Program	6500	6000	5100	5000
OPTION 3: School Immunization Program	650	550	600	500

Number of persons
for whom program options
will increase life expectancy

TABLE B   OUTCOMES FOR ALTERNATIVE HEALTH PROGRAMS

	INCREASE IN POLLUTION		NO INCREASE IN POLLUTION	
	Population Increase	Population Remains Stable	Population Increase	Population Remains Stable
PROBABILITY→	0.25 Moderate	0.35 High	0.10 Low	0.35 High
OPTION 1: Industrial Health Program	5500	5000	4800	4000
OPTION 2: Adult Screening Program	6500	6000	5100	5000
OPTION 3: School Immunization Program	650	550	600	500

Number of persons
for whom program options
will increase life expectancy

FIGURE 5.6 COMPOUND EVENTS (*Explanation:* There are often several types of events that influence the outcomes or payoffs for each action. Table A shows how two types of events, each with two possible conditions, can be compounded to create four mutually exclusive events. Table B shows the same data but structures the events in a hierarchy for greater clarity.)

mutually exclusive. One way is to present each of the events in the matrix without compounding them, but redesigning the matrix so that the multiple possibilities are clearly evident. In the foregoing alternative environmental health programs, the actions could be cross-tabulated against several sets of events labelled population conditions, environmental conditions, economic conditions, organizational conditions, and so on. Each subset of conditions may contain mutually exclusive events, even though the combined grouping of sets of events are not mutually exclusive. A more formal analysis could be computed for each subset, although a conventional decision criterion, similar to those noted previously, could not be used (Figure 5.7).

Events That Are Not Collectively Exhaustive
In addition to the nonexclusivity of events, another common ambiguity exists when the events are not collectively exhaustive, and the analyst cannot account for all the possible situations. Here, the decision matrix can be amended to include event categories like *status quo, conditions remain the same,* or *other unknown events* (Figure 5.8). These categories can be useful in showing the audience the legitimate ambiguity and uncertainty involved in most complex decision-making situations.

The difficulty in using such categories lies in the estimation of outcomes, and may be time-consuming and/or uncertain. If the presentor includes the category *other events*, there must be some accompanying estimate of the range of outcomes associated with each action. Since the events are unknown, the analyst has to present an extraordinarily wide range of possible payoffs or indicate that the payoffs also are unknown. Consequently, this technique should only be used if the *other* category is truly marginal and unlikely to have a real impact on the decision. It should be used only as a device to show the audience that the analysis is as complete as was feasible, given the constraints under which it was prepared.

Ambiguous Payoffs

The payoffs, or outcomes, in a decision matrix can be difficult to measure and compare. Even in complex financial decisions, there are likely to be intangible items and qualitative differences that cannot be accurately expressed in monetary terms. Most of these ambiguities derive from two basic problems in statistical decision theory—the noncomparability of outcomes and the subjective value of outcomes. These difficulties are similar to those discussed in the previous chapter on benefit-cost analysis.

The issue of noncomparability usually occurs as soon as nonmonetary payoffs are involved. For example, two social programs could be evaluated in terms of the number of people they serve, where part of the outcomes would be comparable if measured as the number of *"people"* or numbers of *"service units."* There will still be the issue, however, of the relative financial costs for each program. Combining financial costs and program effects into comparable terms is difficult. As was noted in the previous chapter, it can be achieved by establishing a cost-effectiveness ratio, or by creating a new index (Figure 5.9). Even when this is done, it may still be problematic to present, especially if the analyst has to explain a table with 20 or more abstract cost-effectiveness ratios to the audience.

Another problem connected with noncomparability lies in the combining of probabilities with values such as cost-effectiveness ratios or an abstract index of values. This too can be presented, but obviously can become mystifying to a general audience. If it is already difficult to present and explain effectiveness/cost ratios and decision criteria independently, combining these measures creates a monumental presentation problem. For example, could one easily explain the selection of an action because it had the highest of the lowest possible expected cost-effectiveness ratios? Such a decision criterion

TABLE A   OUTCOMES FOR ALTERNATIVE HEALTH PROGRAMS

	ECONOMIC GROWTH (EMPLOYMENT INCREASE)						NO ECONOMIC GROWTH (EMPLOYMENT REMAINS STABLE)					
	POLLUTION INCREASE			NO POLLUTION INCREASE			POLLUTION INCREASE			NO POLLUTION INCREASE		
	Population Increases	Population is Stable	Population Decrease	Population Increases	Population is Stable	Population Decrease	Population Increases	Population is Stable	Population Decrease	Population Increases	Population is Stable	Population Decrease
PROBABILITY →	.20	.20	.02	.08	.15	.01	.05	.05	.08	.02	.10	.04
OPTION 1: Industrial Health Program	5700	5100	4800	4900	4100	3700	5300	5000	4700	4500	4000	3500
OPTION 2: Adult Screening Program	6600	6100	5800	5200	5100	4700	6400	6000	4800	4600	5000	4500
OPTION 3: School Immunization Program	670	560	530	610	520	450	640	550	520	560	500	410

Number of persons
for whom program options
will increase life expectancy

FIGURE 5.7 MULTIPLE TABLES FOR DIFFERENT TYPES OF EVENTS (**Explanation:** If there are many different types of events, each with several conditions, then a single table is impractical for presentation. Table A shows how one matrix could be stretched to cover 3 types of events and a total of 12 conditional events. Table B shows how with the addition of 1 more conditional event, multiple tables are preferable even though structure of the data seems less precise.)

may be rational, even though it is difficult to explain. The use of several illustrations, computation examples, and explanatory notes is probably the most effective course of action in this instance.

The second major area of difficulty that occurs in measuring outcomes concerns the subjectivity of the values. For example, each member of an audience, such as a board of directors, may attach a different personal value to the same outcome. One decision maker may view a potential loss of program revenues of $100,000 as totally unacceptable (a catastrophe), while another may view it as serious but not unacceptable. Issues such as this have led to complex statistical procedures for estimating the *subjective utility* of different outcomes in abstract terms.

These procedures are useful and may represent the most rational procedure available but are inappropriate when they cannot be communicated effectively to an audience who must comprehend, review, and influence the

TABLE B   OUTCOMES FOR ALTERNATIVE HEALTH PROGRAMS

CRITICAL EVENTS

	Economic Growth	Economic Stability	Economic Decline	Pollution Increases	Pollution does not Increase	Population Increases	Population Remains Stable	Population Decreases
PROBABILITY →	.66	.24	.10	.60	.40	.35	.50	.15
OPTION 1: Industrial Health Program	4900	4700	4300	5300	4500	5100	4600	4200
OPTION 2: Aduct Screening Program	5800	5400	5000	6300	5500	5700	5600	5000
OPTION 3: School Immunization Program	580	550	510	600	530	620	530	480

Number of persons
for whom program options
will increase life expectancy

final decision. Consequently, it is important to present subjective evaluations effectively. One approach would be to include both types of values, combining the objective value (such as monetary figures, or service units or even a cost-effectiveness ratio) along with a subjective measure of its value to the audience members. In this way, it may be possible to evaluate outcomes in ordinal or nominal terms, such as satisfactory, marginal, unacceptable, or excellent (Figure 5.10). By using this technique, several outcomes with different numeric values may appear to have the same ordinal or nonnumeric value. Nevertheless, this approach may be the only way to communicate the issues effectively.

These techniques rely heavily on subjective judgments, but may become the focus of the presentation, incorporating such judgments on behalf of the audience to facilitate the decision- making value of the statistics. The techniques may also prevent the use of more formal computational procedures, but on the other hand can make presentations more effective. If one action or

TABLE A   OUTCOMES FOR ALTERNATIVE HEALTH PROGRAMS

	CRITICAL EVENTS				OTHER EVENTS
	Economic Growth with both Pollution & Population Increase	Economic Growth with Pollution Increase but no Population Increase	Economic Growth but no Population or Pollution Increase	Status Quo (nothing changes)	
PROBABILITIES	.20	.22	.16	.10	Range .01 to .08
OPTION 1: Industrial Health Program	5700	5050	4050	4000	Range 3500 to 5300
OPTION 2: Adult Screening Program	6600	6080	5050	5000	Range 4500 to 6400
OPTION 3: School Immunization Program	670	555	515	500	Range 410 to 640

Number of persons
for whom program options
will increase life expectancy

FIGURE 5.8 UNKNOWN EVENTS *(Explanation:* In some instances the analyst cannot find the information relevant to all possible events. For example, the probabilities or the payoffs may be unknown. This table shows how such a situation could be presented. It presumes that the analyst can find the relevant data for only some of the events shown in Figure 5.7 (Table A) and that the last column represents a rough estimate.)

program leads to subjective evaluations of *satisfactory* for all possible events while another is rated *excellent* in some cases and *marginal* in others, a decision by numbers might favor the latter program, whereas a decision based on subjective labels might favor the former. Which approach is more rational will depend on both the problem at hand and the audience being addressed.

A more difficult approach to presenting subjective judgments tries to convert them to a precise numeric evaluation. As was stated previously, there are algorithms for doing this in which the outcome is labelled with a number, which is explained as a complex estimation of relative trade-offs among various outcomes. This procedure can be accurate and useful, although the most effective presentation includes a clear hypothetical illustration supplemented with one or two examples of how specific values in the presentation have been derived (Figure 5.11).

Audience Interests versus Events

A radically different approach to cross-tabulating actions against future events is to cross-tabulate them against various segments of audience interests. This is analogous to the use of coalitions, or sets of interests, in presenting cost-benefit and cost-effectiveness data. One side of the matrix lists the actions, while the other dimension is oriented directly to the sets of interests in the audience as determined by the presentor. For example, several job development programs could be cross-tabulated against different communities, types of workers or age groups, each of which represents different audience interests. Two of the job programs have the same net effect of creating new jobs, but the geographic distribution of those jobs is different, as one favors community *A* and the other favors community *B*. It is a disservice to the audience if this information is relevant and the presentation hides or obscures the issue. The intention is to help the audience by aggregating beforehand the outcomes in which they are most interested.

In labelling each set of interests, the presentor can use the names of specific political entities (such as local units of government), or alternatively focus on some other key variable. For example, if members of the audience are indirectly representing different segments of the work force, the labels could name the various types of workers involved, such as skilled workers, unskilled workers, minority groups, suburban areas, and so on. The labels need not be mutually exclusive and/or collectively exhaustive. The intent is to be relevant to the audience interests, not to create a table with a clear parallel structure in the subtitles (Figure 5.12).

In some instances, decision criteria can be applied, although rarely with the rigor of a conventional matrix with monetary payoffs and precise probabilities. Thus, it may be possible to show that the outcomes for one action have satisfactory results for almost all the principal constituencies in the audience, whereas the outcomes for competing actions are excellent for some interests but unsatisfactory for others. This type of presentation and analysis may lead to far different conclusions than more traditional approaches in statistical decision theory. It is discussed here principally because it is one of the few techniques for decision making that is integrated directly with the goal of effective presentation to an audience.

DECISION HIERARCHIES: MULTIPLE ACTIONS, EVENTS, AND OUTCOMES

Decision trees or hierarchies offer a more complex but powerful technique for presenting decision-making data. The principal advantage is that decision

TABLE A   RECOMMENDED ACTION FOR HEALTH PROGRAM

	CRITICAL EVENTS				OTHER EVENTS
	Economic Growth with both Pollution & Population Increase	Economic Growth with Pollution Increase but no Population Increase	Economic Growth but no Population or Pollution Increase	Status Quo (nothing changes)	
PROBABILITIES →	Moderate	Moderate	Moderate	Low	Low
OPTION 1: Industrial Health Program Rational: Option 1 is recommended because it offers the lowest cost per effective treatment in almost all cases.	$559	$486	$543	$660	$498 to $685
OPTION 2: Adult Screening Program	$528	$576	$693	$700	$656 to $866
OPTION 3: School Immunization Program	$746	$721	$776	$800	$687 to $939

Cost per person for each program option
(for persons whose life span will be increased)

FIGURE 5.9 MEASURING PAYOFFS (*Explanation:* Traditionally payoffs for business decisions are measured in dollars. However, as is shown in Figure 5.6 to 5.8 there are other ways to quantify payoffs. One approach is to use effectiveness/cost ratios or benefit/cost ratios. Table A does this for the data in Figure 5.8 and also illustrates a presentation for a recommended decision. Table B does the same for the data in Figure 5.2.)

hierarchies can portray a series of contingent actions, events, and outcomes. They can also incorporate more sophisticated use of probabilities, decision criteria, subjective evaluation, and noncomparable and qualitative values. These factors are extraordinarily difficult to communicate effectively to an audience that is ignorant of the underlying computations, methods, and theories. Nevertheless, decision hierarchies and trees can be useful if they are presented as a decision-making aid rather than as a complete algorithm leading to a final conclusion. This section does not address formal theoretical and computational procedures but instead the informal, improvisational use of these techniques as a presentation device.

TABLE B   RECOMMENTATION FOR CLINIC CONSTRUCTION

	ENROLLMENT PROJECTIONS		
	MODERATE ENROLLMENT	HIGH ENROLLMENT	LOW ENROLLMENT
PROBABILITY ──▶	0.50	0.30	.020
[A.] NEW SATELLITE FACILITIES COST = $400,000 Recommended Action: The high benefit/ cost ratios for the most likely events clearly outweigh the poor ratio for the least likely event.	Actual (Expected) Benefit $6,000,000 (3,000,000) B/C Ratio = 7.5	Actual (Expected) Benefit $12,000,000 (3,600,000) B/C Ratio = 9.0	Actual (Expected) Benefit -$3,000,000 (-600,000) B/C Ratio = -1.5
[B.] STAGED SATELLITE DEVELOPMENT COST = $350,000	$3,200,000 (1,600,000) B/C Ratio = 4.57	$7,500,000 (2,250,000) B/C Ratio = 6.43	-$600,000 (-120,000) B/C Ratio = -0.34
[C.] NO NEW CONSTRUCTION COST = $100,000	$1,000,000 (500,000) B/C Ratio = 5.0	-$500,000 (150,000) B/C Ratio = -1.5	$500,000 (100,000) B/C Ratio = 1.0
[D.] SINGLE NEW MAJOR FACILITY COST = $700,00	-$2,000,000 (-1,000,000) B/C Ratio = -1.43	$14,000,000 (4,200,000) B/C Ratio = 6.0	-$4,000,000 (-800,000) B/C Ratio = -1.14

NOTE: B/C Ratio = (Expected Benefit/ Cost)

Actions versus Events

A basic presentation issue lies in making the distinction between the decision made based on alternative actions and the events which may occur subsequent to those actions. Typically, this is achieved with graphic symbols located at each *node* or *branching point* in the tree diagram. A square shaped node, for example, can represent actions or decisions, while a circle can represent events. A decision hierarchy might begin with three alternative actions branching out from a decision point, shown as a square. Each action may signify a different type of program, and each action or program leads to another node. These nodes could be other decisions, such as optional program modifications, which may vary from one program to another and would also be shown as squares (Figure 5.13). Alternatively, the nodes could be shown as circles that represent *chance* points where several different events might occur. For example, if an economic development program were being considered, the events emanating from the chance node might represent alternative economic conditions beyond the control of the decision makers (Figure 5.14).

CRITICAL EVENTS

PROBABILITY	Economic Growth	Economic Stability	Economic Decline	Pollution Increases	Pollution does not Increase	Pollution Increases	Population Remains Stable	Population Decreases
	HIGH	LOW	LOW	HIGH	MOD-ERATE	MOD-ERATE	MOD-ERATE	LOW
OPTION 1: Industrial Health Program	S	S	U	G	S	G	S	U
OPTION 2: Adult Screening Program	G	S	S	G	S	G	S	S
OPTION 3: School Immunization Program	S	S	U	S	S	S	S	U

Relative value of outcomes based on costs, number of people treated and effect of treatment on quality of life and economic productivity.

KEY: G = GOOD; S = SATISFACTORY; U = UNSATISFACTORY

FIGURE 5.10 QUALITATIVE ORDINAL EVALUATIONS IN A DECISION MATRIX **(Explanation:** In some circumstances it may be more effective, albeit less precise, to present decision payoffs in qualitative terms. This example parallels the information shown in Figure 5.7.)

The various lines or nodes are each labelled with appropriate statistical data, so that each branch after a chance node might be labelled with the probability that it will occur. This may be a precise number or ordinal statement such as high, medium, or low probability. It should be emphasized that the use of precise probabilities is complex, and should follow the necessary theoretical constraints. For instance, the sum of the probabilities after each chance node should equal 1.0 *if* the events are mutually exclusive, collective exhaustive, *and* conditional on the occurrence of the sequence of events leading to that chance point.

Other data presented may include costs, benefits, and effects associated with taking actions, the occurrence of events, and so on. In addition, it is important to label actions and events in as finite a manner as possible. Audiences may find decision hierarchies too abstract, artificial, or otherwise unrepresentative of what they personally perceive to be the decision-making process. Clarifying the information can, to some extent, alleviate this problem.

It is possible that the same actions or events may occur more than once in a diagram. For example, alternative economic conditions may be a set of

TABLE A   RECOMMENDATION FOR HEALTH PROGRAM DEVELOPMENT

		CRITICAL EVENTS				OTHER EVENTS
		Economic Growth with both Pollution & Population Increase	Economic Growth with Pollution Increase but no Population Increase	Economic Growth but no Population or Pollution Increase	Status Quo (nothing changes)	
	PROBABILITIES →	0.20	0.22	0.16	0.10	Range .01 to .08
OPTION 1: Industrial Health Program	Actual Points	7.75	7.50	6.25	5.25	4.25 to 6.80
	Expected "Point" Value	1.55	1.65	1.00	0.53	0.20 to 0.60
OPTION 2: Adult Screening Program	Actual Points	8.00	7.00	6.25	6.00	2.50 to 4.75
	Expected "Point" Value	1.60	1.54	1.00	0.60	0.25 to 0.75
OPTION 3: School Immunization Program	Actual Points	8.50	7.50	6.00	5.00	3.00 to 4.75
	Expected "Point" Value	1.70	1.65	0.96	0.50	0.25 to 0.85

Recommended Action
Option 3 offers the highest expected valuesfor the most probable events, based upon the judgement of the review panel

Point scores are the average of four combined ranks based on [1] problem cost, [2] number of persons treated, [3] improvement in quality of life and [4] increased economic productivity. Highest possilbe score equals 1.0. All factors were weighted equally.

FIGURE 5.11 SUBJECTIVE QUANTITATIVE EVALUATIONS IN A DECISION MATRIX (**Explanation:** There are various techniques for assigning subjective quantitative values to the outcomes. For example, an expert panel can assign points to each attribute or feature of an outcome, and the points then can be averaged. Tables A and B show how such a system might be presented. Table A is based on the situations shown in Figures 5.8 and 5.9. Table B is based on the data found in Figures 5.2 and 5.9.)

events that is displayed in the diagram several times. That is, after each alternative action, there will be a chance that each of the alternative economic events may occur (Figure 5.15). Similarly, an action (such as expansion of the economic development program into a new geographic area) might be an alternative that occurs at several points in the decision-making hierarchy. If the repetition of events and actions becomes too visually cumbersome, it may confuse or otherwise detract from the presentation. Under these circumstances, the use of a decision matrix may be a preferable presentation technique. The principal advantage of the decision tree is that, unlike the matrix,

TABLE B   RECOMMENDATION FOR CLINIC DEVELOPMENT

HEALTH PLAN ENROLLMENTS

		Moderate Enrollment	High Enrollment	Low Enrollment
	PROBABILITY →	0.50	0.30	0.20
[A] No New Costruction*	Actual Scores	10.00	9.00	9.50
	Expected Point Value	5.00	2.70	1.90
[B] Single New Major Facility	Actual Scores	2.90	4.80	1.90
	Expected Point Value	1.45	1.44	0.38
[C] New Satellite Facilities	Actual Scores	6.40	6.20	3.90
	Expected Point Value	3.20	1.26	0.78
[D] Staged Satellite Development	Actual Scores	8.20	8.60	6.60
	Expected Point Value	4.10	2.58	1.32

* At this time no new construction is recommended. This is due to the relatively high construction costs of other options, the importance of minimizing those costs and the fact that without new construction there are still positive benefits.

The point scores represent a weighted average of three factors: The cost of each option (weight = 50%), the potential financial benefit (30%) and the estimated degree of difficulty in implementation (20%). These weights were applied to the numerical rank order for each possibility. The best possible value was a score of 12.0. The worst value was 1.0. For example, the scores in the lower right corner were based on a rank of 8.0 for cost, a rank of 4 for financial benefit and a rank of 7 for organizational difficulty. The final actual score of 6.6 = (50% x 8) + (30% x 4) + (20% x 7). The expected point value = 6.60 x 0.20 = 1.32.

TABLE A   DECISION MAKING EVALUATION BY GROUP

	RELEVANT GROUPS							
	Industry & Manufacturing Sector		Other Employees		City Neighborhoods		Rural & Suburban Areas	
	Population Increases	Population is Stable	Population Increases	Population is Stable	Population Increases	Population is Stable	Population Increases	Population is Stable
PROBABILITY ➔	0.40	0.60	0.40	0.60	0.40	0.60	0.40	0.60
OPTION 1: Industrial Health Program	5	5	3	2	4	4	2	1
OPTION 2: Adult Screening Program. Recommended Action: This option is at least satisfactory for three of the decision-making groups regardless of population conditions. For the fourth group (rural & suburban) it is as good or better than the other two options.	4	3	4	3	5	5	3	1
OPTION 3: School Immunization Program	2	2	3	2	5	4	3	1

VALUES: **5** -excellent
             4 -good
             3 -satisfactory
             2 -poor
             1 -unsatisfactory

FIGURE 5.12 MATRIX FOR DECISION-MAKING GROUPS (**Explanation:** Different types of decision makers may be concerned with different factors for making a decision. In such situations the analysis should clearly present different factors that are relevant to each group and, if possible, should make distinct recommendations for each group. This example is based on a situation similar to that presented in Figure 5.6.)

it can easily portray a linear sequence of conditional actions and events. Both techniques have shortcomings, and the final choice will depend on the type of information to be presented as well as the nature of the audience.

Outcomes and Payoffs

Final outcomes and payoffs traditionally are shown at the end points of the hierarchy, or the tips of the tree branches. Each outcome is associated with a particular trail of actions and events, and some outcome measures may be

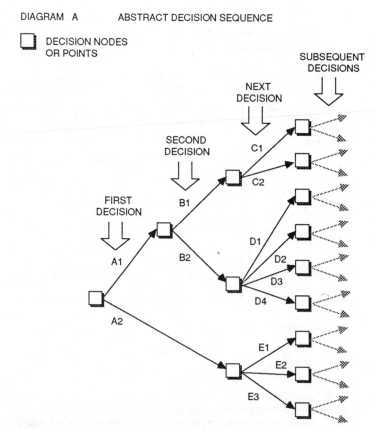

FIGURE 5.13 A SIMPLE DECISION TREE (***Explanation:*** Many decision-making situations involve complex combinations of choices. These can be shown in a simple hierarchical or tree diagram. Diagrams A and B illustrate decision-making situations similar to those shown in Figure 5.1.)

listed along the chain of actions and events. A conventional approach is to show the cost of each action at the point at which it occurs, and then show the final set of benefits at the end of each sequence. In practice, however, there may be a series of costs, benefits, and effects at several points. Showing outcomes adjacent to the actions or events to which they are related as well as at the tips of the tree is probably the most effective technique (Figure 5.16).

As with decision matrices, cost-benefit analyses and cost-effectiveness analysis, there are many difficulties presenting noncomparable and qualitative outcomes. It is possible to use effectiveness/cost ratios, indices or values

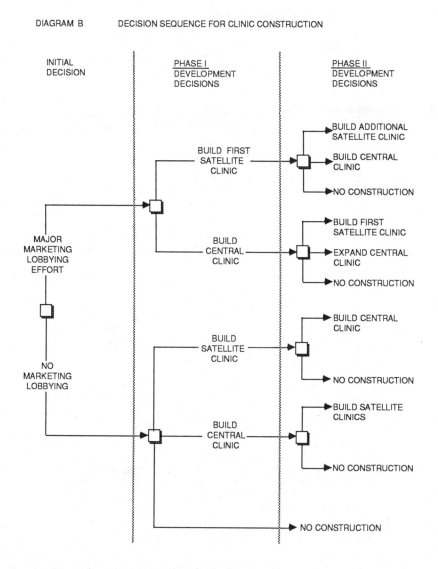

DIAGRAM B   DECISION SEQUENCE FOR CLINIC CONSTRUCTION

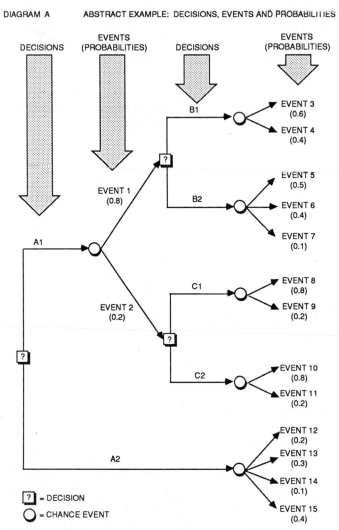

DIAGRAM A   ABSTRACT EXAMPLE: DECISIONS, EVENTS AND PROBABILITIES

FIGURE 5.14   DECISION TREES WITH CHANCE POINTS   (*Explanation:* Frequently, sequential decisions must be based on events that have different probabilities of occurring. Diagram A is an abstract example, and Diagram B is a concrete example. There are complex mathematical rules underlying the computation of these probabilities. For example, in Diagram A the probabilities of Events 5, 6, and 7 sum to 1.0 because they are mutually exclusive and collectively exhaustive. But this is only the case *after* Event 1 has occurred *and* action B2 is selected. That is, the *conditional* probability of Event 6 is .4 given Event 1 and B2. The *combined* probability, however, that both Event 1 and 6 will occur (if decision B2 is chosen) is .8 x .4 or .32.)

DIAGRAM B   CONCRETE EXAMPLE: DECISIONS, EVENTS AND PROBABILITIES

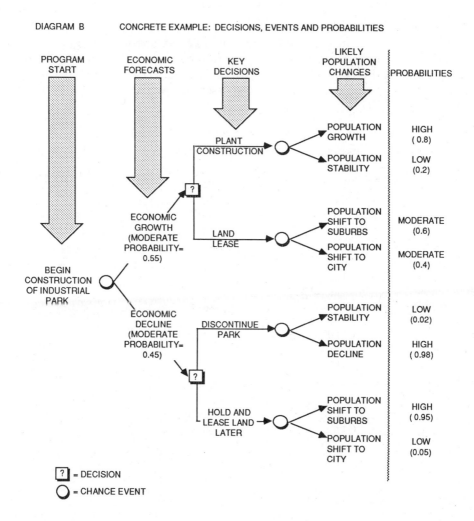

☐? = DECISION

◯ = CHANCE EVENT

or other similar measures in a decision tree diagram. There are, however, limits to an audience's ability to comprehend such measures. Consequently, it may be appropriate to try to scale down or limit the length and complexity of a decision tree diagram if more complex measures or outcomes are to be presented (Figure 5.17).

Outcomes shown in a decision tree diagram are never truly final outcomes, as decision trees presume that actions and events are sequenced over time. Therefore, the presentor must decide at what point in time to end the trail of events, actions, and outcomes. This issue is analogous to the problem of presenting second order and third order costs and benefits discussed in the previous chapter. In the case of the economic development programs, for example, the outcomes could end with the number of new jobs that are created. Alternatively, the chain of events and decisions could be continued to display how new jobs would effect other businesses and public services within the area.

DECISION MAKING SITUATION FOR NEW ECONOMIC DEVELOPMENT PROGRAM

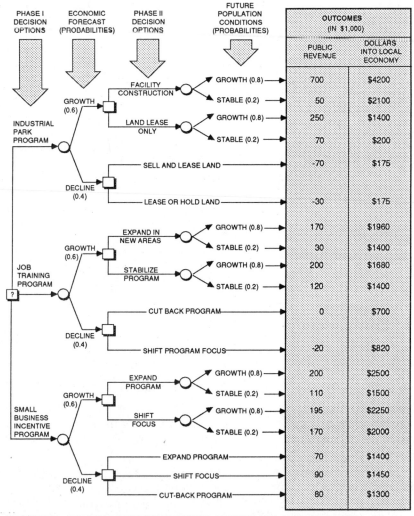

FIGURE 5.15 COMBINING DECISIONS, EVENTS, PROBABILITIES, AND OUTCOMES (*Explanation:* The decision tree becomes relevant to the audience of decision makers when they see the potential payoffs. The simplest way to show payoffs is at the tips of the branches. This diagram expands the decision-making situation shown in Figure 5.14.)

One presentation technique gives visual emphasis to the point in the sequence at which the most important outcomes occur. For example, dashed or dotted lines could continue the chain past the primary outcomes (Figure 5.18). An alternative strategy would be to list subsequent events, decisions, and outcomes after the primary outcomes in a simple chart format, without any additional lines or nodes. Again, the choice of the presentation strategy depends on the relative importance of nonprimary events to the audience.

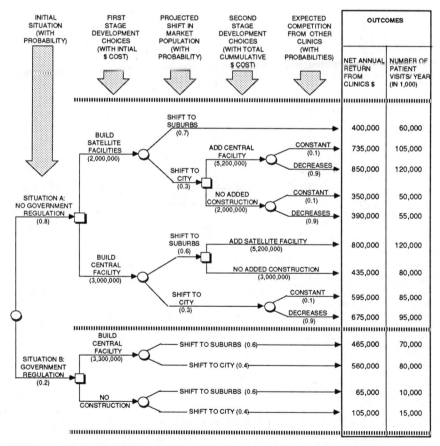

FIGURE 5.16 PRESENTING INFORMATION AT NODES (*Explanation:* Decision trees allow for the presentation of information at nodal points as well as at the "tips" of the tree. This diagram, based on the situation shown in Figure 5.13, indicates the costs of each action along the way. It also illustrates how a decision-making situation can begin with an event or chance point rather than a decision or choice point.)

OPTIONS AND OUTCOMES FOR ECONOMIC DEVELOPMENT

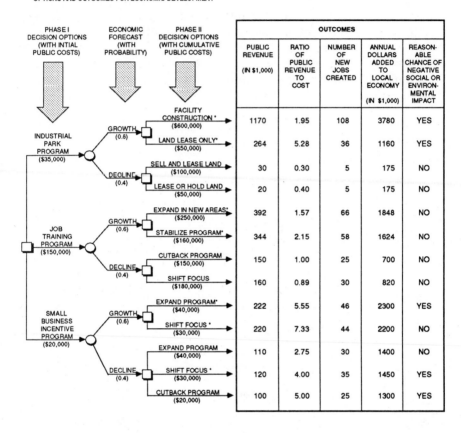

PHASE I DECISION OPTIONS (WITH INITIAL PUBLIC COSTS)	ECONOMIC FORECAST (WITH PROBABILITY)	PHASE II DECISION OPTIONS (WITH CUMULATIVE PUBLIC COSTS)	PUBLIC REVENUE (IN $1,000)	RATIO OF PUBLIC REVENUE TO COST	NUMBER OF NEW JOBS CREATED	ANNUAL DOLLARS ADDED TO LOCAL ECONOMY (IN $1,000)	REASONABLE CHANCE OF NEGATIVE SOCIAL OR ENVIRONMENTAL IMPACT
		FACILITY CONSTRUCTION * ($600,000)	1170	1.95	108	3780	YES
INDUSTRIAL PARK PROGRAM ($35,000)	GROWTH (0.6)	LAND LEASE ONLY* ($50,000)	264	5.28	36	1160	YES
	DECLINE (0.4)	SELL AND LEASE LAND ($100,000)	30	0.30	5	175	NO
		LEASE OR HOLD LAND ($50,000)	20	0.40	5	175	NO
JOB TRAINING PROGRAM ($150,000)	GROWTH (0.6)	EXPAND IN NEW AREAS* ($250,000)	392	1.57	66	1848	NO
		STABILIZE PROGRAM* ($160,000)	344	2.15	58	1624	NO
	DECLINE (0.4)	CUTBACK PROGRAM ($150,000)	150	1.00	25	700	NO
		SHIFT FOCUS ($180,000)	160	0.89	30	820	NO
SMALL BUSINESS INCENTIVE PROGRAM ($20,000)	GROWTH (0.6)	EXPAND PROGRAM* ($40,000)	222	5.55	46	2300	YES
		SHIFT FOCUS * ($30,000)	220	7.33	44	2200	NO
	DECLINE (0.4)	EXPAND PROGRAM ($40,000)	110	2.75	30	1400	NO
		SHIFT FOCUS * ($30,000)	120	4.00	35	1450	YES
		CUTBACK PROGRAM ($20,000)	100	5.00	25	1300	YES

* The outcomes associated with these decisions will vary with changes in the population. The numbers shown are an expected value based upon an 80% chance of population growth and a 20% chance of population stability.

FIGURE 5.17 PRESENTING A VARIETY OF OUTCOMES (*Explanation:* The payoffs for different sequences of choices and events may include a variety of measures such as dollar figures, abstract ratios, or qualitative events. This figure presents information for the situation shown in Figure 5.15. However, because of the complexity of the added information, the number of "branches" in the tree have been reduced by eliminating the chance events associated with the population changes shown in Figure 5.15 and simply referencing these as footnotes.)

Decision Criteria

There are established computational methods and constraints for determining the most rational course of action displayed on a decision tree. Typically, these procedures begin with collapsing or combining the values of outcomes at the tips or end of the hierarchy. The evaluation process works backward from the

CLINIC CONSTRUCTION PROGRAM:
PRIMARY OUTCOMES AND SECONDARY EFFECTS

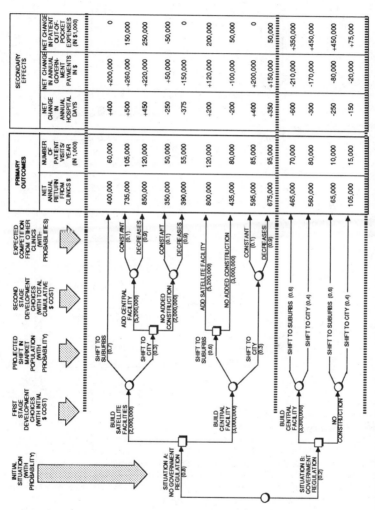

Figure 5.18 Presenting Secondary and Subsequent Outcomes (*Explanation:* In addition to direct costs, benefits, and effects there may be a variety of secondary or tertiary outcomes that should be noted. These can be presented in table format and deemphasized. This example expands the data shown in Figure 5.16.

TABLE A   RECOMMENDED DECISION STRATEGY BASED ON HIGHEST EXPECTED PUBLIC REVENUE

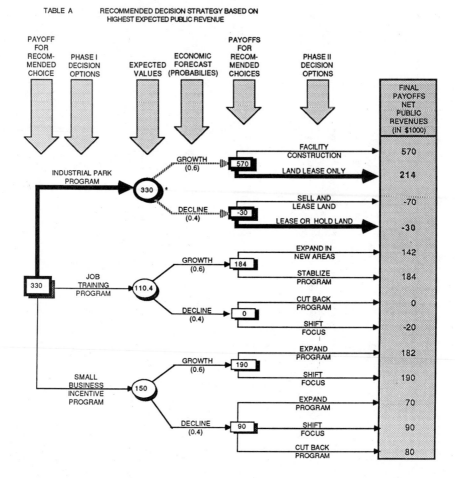

* EXAMPLE COMPUTATION: $ 330 = 0.6 (570) + 0.4 (-30)

FIGURE 5.19 APPLYING EXPECTED VALUE CRITERIA **(*Explanation:* When presenting a decision hierarchy it is customary to indicate what criteria are to be used in making decisions and how they can be applied. This example shows the application of an *expected value* criterion to the data shown in Figure 5.17.

      Theories and procedures for computing these values can be found in other texts. Briefly, the expected value at each chance point is calculated by starting at the tips and multiplying each probability by the *value* of the decision. Then all these values are added together at the associated chance point. The expected value shown at each decision point is the highest of the alternative values shown at the next node to the right—there are no computations required, just an inspection of the data. In this hierarchy, two sets of expected values are shown—one based on net public costs (Table A), and one on revenue/cost ratios (Table B). The right choice of a criterion depends on the analysts' judgment and the audience.)

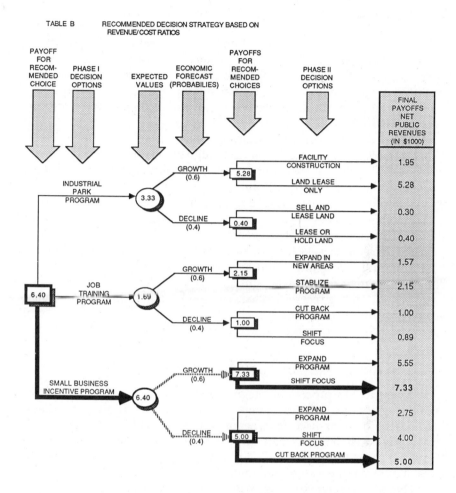

TABLE B   RECOMMENDED DECISION STRATEGY BASED ON
               REVENUE/COST RATIOS

tips of the tree chain toward the beginning, and the values of the last events are combined, generating a value for the last decision. These values are then combined to find the value of the next to the last decision. The process continues until there is a value attached to each of the alternative actions that emanate from the first decision point at the beginning of the tree or hierarchy.

It is usual to select the initial action with the best value, but the process can be defined to pick the action with the highest expected value, to minimize the maximum expected loss, or to maximize the maximum gain (Figure 5.19). The choice of the decision criteria depends on the judgment of the person or group with the authority to make the decision, and procedures for computing and applying these rules can be found in other texts (see Appendix for references).

The presentation problem here is how to explain the decision criteria to the audience. At the very least, the alternative chain of recommended decisions should be given clear visual emphasis. Sometimes, visual emphasis should also be given to other values and outcomes that are key to the selection process. If the decision criterion is to select a program that avoids major negative consequences, the presentor should highlight all such negative outcomes and indicate the effectiveness of the recommended decision sequence (Figure 5.20).

One of the advantages of a decision tree lies in its ability to focus attention on the immediate decision problem. Although the chain of events, actions, and outcomes extends into the future, the evaluation process always leads the audience back to the starting point—the immediate, or first, decision. This should also be given clear visual emphasis in the presentation.

Another advantage of a decision hierarchy is that it is possible to portray the use of different criteria at different points in the sequence. For example, the primary decision of selecting an economic development program may be based on minimizing maximum loss and avoiding the worst possible outcomes. There may, however, be subsequent choices that modify or expand the basic program being recommended. These subsequent choices could be presented in terms of other criteria, such as maximizing expected value or aiming for the best cost-effectiveness ratio (Figure 5.21).

Decision Hierarchies for Audience Interests

If the presentation process is integral to decision making, it is essential to express the situation in a clear, comprehensible manner. The decision diagram becomes the focus for discussion, negotiation, and debate. As with the decision matrix, the decision tree or hierarchy can be molded to correspond directly to the audience, which may be made up of political constituencies or representatives of different theories or opinions.

One way to achieve this is to divide the outcome measures at the end of the diagram into several branches, each representing a different set of interests. If the analyst has prepared a decision tree for selecting an economic development program, the tips could be branched one step further to show more discrete outcomes for different types of workers, communities, and/or industries (Figure 5.22).

A variation of this technique lies in the construction of a table using the end points of the tree as the vertical dimension (left side) of a table. Across the top, the table lists different sets of interests. The cells describe the interrelation between each principal outcome and each set of audience interests (Figure 5.23).

In some cases, it may be appropriate to show discrete outcomes at several nodes along the chain of actions and events. This entails enlarging the relevant decision nodes so that each one can contain more data (Figure 5.24). If this

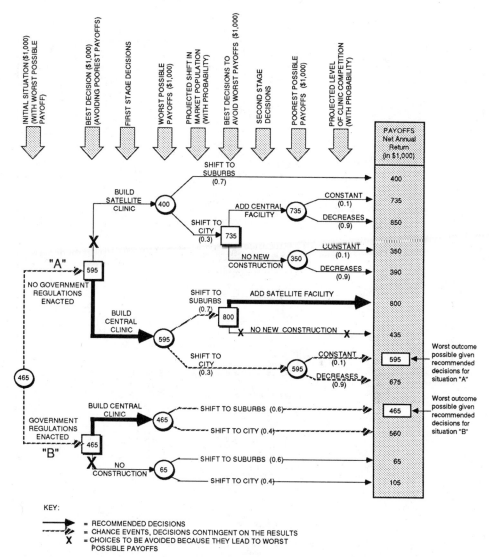

RECOMMENDED DECISION STRATEGY FOR AVOIDING
POOREST PAYOFFS

KEY:

= RECOMMENDED DECISIONS
= CHANCE EVENTS, DECISIONS CONTINGENT ON THE RESULTS
X = CHOICES TO BE AVOIDED BECAUSE THEY LEAD TO WORST
  POSSIBLE PAYOFFS

FIGURE 5.20 APPLYING OTHER CRITERIA (**Explanation:** In addition to expected values, other common criteria include minimizing the worst possible outcomes. In the decision matrix in this example, based on Figure 5.18, there are no added computations. The only difference is emphasizing and presenting the decision sequence. The sequence is determined by working backward from the ends of the tree as follows: for each chance point the worst outcome is presumed to occur, and it is indicated on the diagram; at each decision point the branch is taken that avoids the worst subsequent outcomes. The use of these criteria with other payoffs as a basis for the decision would lead to different conclusions.)

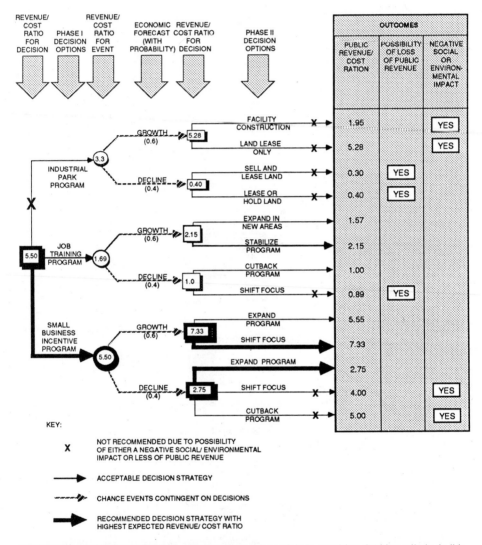

FIGURE 5.21 COMBINING CRITERIA (***Explanation:*** It is possible to combine decision criteria. In this example, based on data from Figure 5.17 and 5.19, one criterion is to avoid (minimize) negative impacts. Those options are noted and "crossed off" the hierarchy. For the remaining options a criterion of maximizing the expected values is applied. The appropriate combination of criteria depends on the analyst and intended audience.)

DECISION-MAKING ISSUES FOR LOCAL GOVERNMENT, LABOR AND BUSINESS

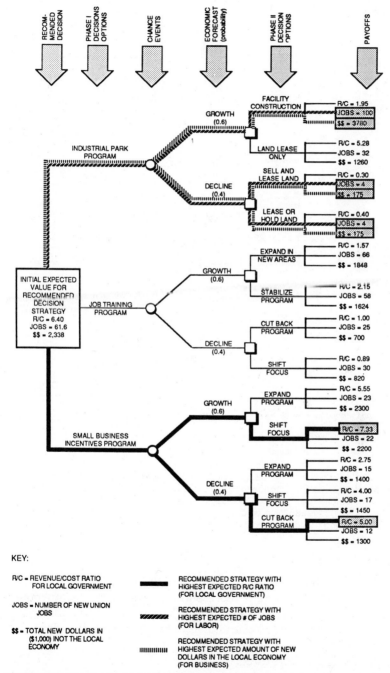

FIGURE 5.22 MULTIPLE PAYOFFS AND CRITERIA (*Explanation:* In some instances several payoffs can be shown simultaneously, each responding to a different decision-making audience. In this diagram, based on Figure 5.17, three audiences are identified and three decision strategies are shown. The added complexity of showing three strategies may require showing less detail than in Figure 5.17. Also, the criteria for each strategy need not be identical.)

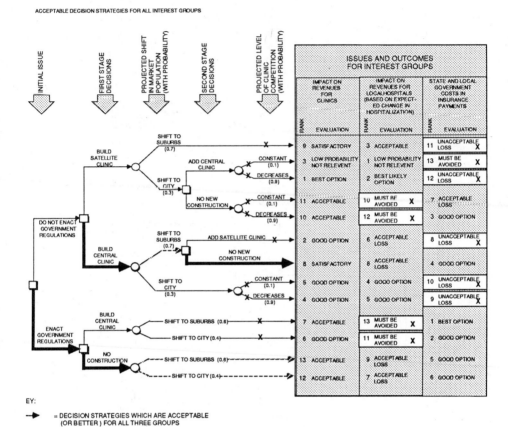

FIGURE 5.23 REPRESENTING DIFFERENT INTERESTS IN DECISIONS (*Explanation:* As the amount and the complexity of data increase, the presentation should focus more on the issues. For example, this figure summarizes some of the issues noted in Figures 5.16, 5.18, and 5.20 and presents them as they relate to three interest groups. Precise data on costs, revenues, and quantitative effects have been replaced by ordinal ranks and qualitative evaluations.)

becomes too cumbersome, an alternative would be to construct more than one decision tree where each diagram represents the evaluation of outcomes attributed to one coalition of interests (Figure 5.25). This, too, can be confusing to present if the diagrams are complicated, or if there are too many diagrams that have to be shown.

There is, however, a more direct way to integrate audience interests. The techniques noted previously treat audience interests primarily as additional notations. In some cases, audience interests can be reflected in the definition of

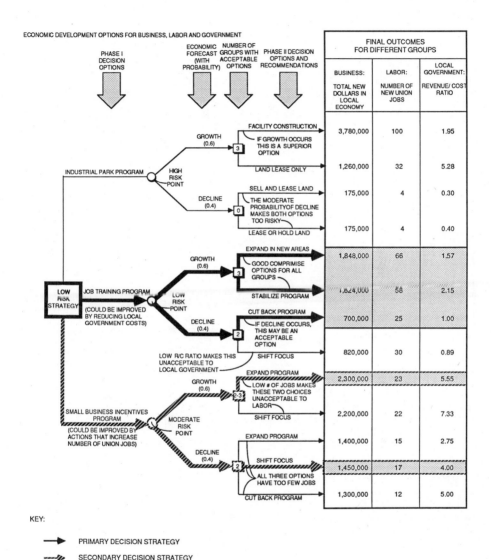

FIGURE 5.24 ANNOTATED DECISION ANALYSIS (Explanation: Another approach to presenting multiple outcomes for divergent audiences is to annotate the elements of the diagram with summary statements. This illustration presents the same decision strategies as in Figure 5.22, but with different analystic assumptions, conclusions, and techniques.)

TABLE A   DECISION STRATEGIES AND ISSUES FOR CLINIC BOARD OF DIRECTORS

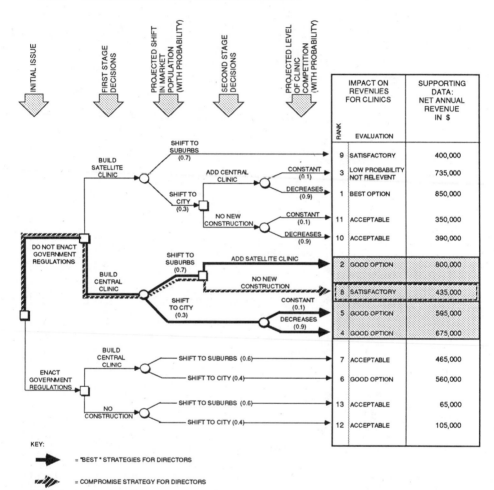

FIGURE 5.25   MULTIPLE DIAGRAMS FOR MULTIPLE INTERESTS   (*Explanation:* To show more information, single diagrams can be disaggregated into mulitple diagrams. This example is based on Figure 5.23. It presumes that the audience needs more than the general information presented in Figure 5.23, including more detailed payoffs and decision-making criteria. The presentation is divided into three parts—Tables A, B, and C, each relating to a different interest group with conflicting values that may, or may not be resolvable in a single decision-making strategy.)

alternative actions and events. In the example of the economic development program, there may be specific actions that could be taken by different communities, industries, unions, or other organizations. If these actions have significant impacts and relate directly to members of the audience, they can be represented as integral or continuing parts of the decision tree (Figure 5.26).

TABLE B DECISION STRATEGIES AND ISSUES FOR HOSPTIAL COUNCIL

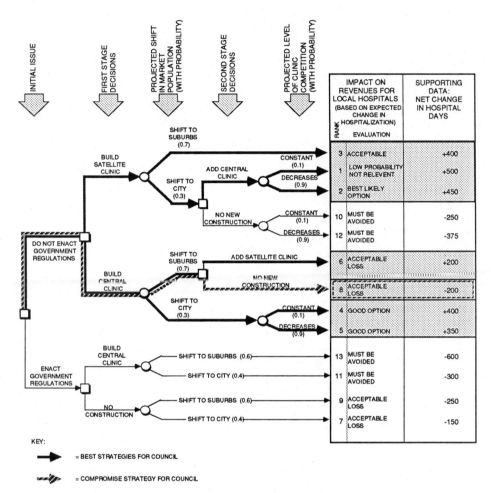

KEY:

➡ = BEST STRATEGIES FOR COUNCIL

⟿ = COMPROMISE STRATEGY FOR COUNCIL

There will then be an automatic need to describe subsequent events, actions and outcomes tailored to audience interests.

This last technique—structuring alternative decisions according to audience interests—may radically alter a decision hierarchy. The traditional decision tree presumes that only a single choice or action can emanate from a decision point. By including, in one diagram, decisions by several groups, the presentor is in effect creating nested decision hierarchies that may be independent of each other. More than one set of alternative actions may emanate from a decision point, and the choice of one subsequent action by one group can be made regardless of a choice of action by another.

This scenario undoubtedly could be redefined to produce a set of actions and events that fit the logical constraints of statistical decision theory. If four groups each choose one of three actions and each choice is independent of the other, the scenario could be expressed as 12 possible compound events. Each event represents a combination of actions, and each has its own probability of occurrence. This may, in fact, be the more rational analytic process, but it usually is not the most effective presentation of the issue.

Again, the presentor must make trade-offs between following the logical constraints of a conventional decision tree methodology and modifying such techniques to help the audience.

TABLE C   DECISION STRATEGIES AND ISSUES FOR STATE GOVERNMENT

Column headers (left to right): INITIAL ISSUE | FIRST STAGE DECISIONS | PROJECTED SHIFT IN MARKET POPULATION (WITH PROBABILITY) | SECOND STAGE DECISIONS | PROJECTED LEVEL OF CLINIC COMPETITION (WITH PROBABILITY)

RANK	STATE AND LOCAL GOVERNMENT COSTS IN INSURANCE PAYMENTS — EVALUATION	SUPPORTING DATA: NET CHANGE IN GOVERNMENTAL PAYMENTS
11	UNACCEPTABLE LOSS	+200,000
13	MUST BE AVOIDED	+260,000
12	UNACCEPTABLE LOSS	+220,000
7	ACCEPTABLE LOSS	+50,000
3	GOOD OPTION	-150,000
8	UNACCEPTABLE LOSS	+120,000
4	GOOD OPTION	-100,000
10	UNACCEPTABLE LOSS	+200,000
9	UNACCEPTABLE LOSS	+150,000
1	BEST OPTION	-210,000
2	GOOD OPTION	-170,000
5	GOOD OPTION	-80,000
6	GOOD OPTION	-20,000

Decision tree branches:

DO NOT ENACT GOVERNMENT REGULATIONS
- BUILD SATELLITE CLINIC
  - SHIFT TO SUBURBS (0.7) → (rank 11)
  - SHIFT TO CITY (0.3)
    - ADD CENTRAL CLINIC
      - CONSTANT (0.1) → (rank 13)
      - DECREASES (0.9) → (rank 12)
    - NO NEW CONSTRUCTION
      - CONSTANT (0.1) → (rank 7)
      - DECREASES (0.9) → (rank 3)
- BUILD CENTRAL CLINIC
  - SHIFT TO SUBURBS (0.7)
    - ADD SATELLITE CLINIC → (rank 8)
    - NO NEW CONSTRUCTION → (rank 4)
  - SHIFT TO CITY (0.3)
    - CONSTANT (0.1) → (rank 10)
    - DECREASES (0.9) → (rank 9)

ENACT GOVERNMENT REGULATIONS
- BUILD CENTRAL CLINIC
  - SHIFT TO SUBURBS (0.6) → (rank 1)
  - SHIFT TO CITY (0.4) → (rank 2)
- NO CONSTRUCTION
  - SHIFT TO SUBURBS (0.6) → (rank 5)
  - SHIFT TO CITY (0.4) → (rank 6)

KEY:
→ = BEST STRATEGIES FOR STATE
⇢ = COMPROMISE STRATEGY FOR STATE

RANKING AND WEIGHTING

The relative evaluation of noncomparable and qualitative costs, benefits, and effects is a recurring problem. At times, this is the crux of an analytic decision-making dilemma, and several techniques have been developed for comparing, ranking, and assigning numerical weights to various outcomes. A common procedure is the ranking and weighting of objectives to determine relative priorities. This approach can be extended to alternative program outcomes, and is especially useful when there are several inseparable components.

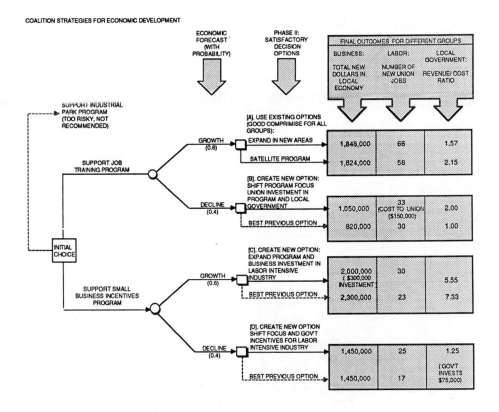

FIGURE 5.26 SUMMARIZING AND BALANCING AUDIENCE INTERESTS (*Explanation:* The most useful statistics are those that summarize the key issues and present only the most relevant data to different audience interests. This example presumes the analyst has presented Figure 5.24 and subsequently wishes to present some "new" decision options that imply new actions on the part of the three interest groups noted in the tabulation of the outcomes. Options B, C, and D each indicate how actions by one of the groups can affect the outcomes and thereby change the decision strategy selected. No single strategy is recommended. Instead, the audience is presented with a series of satisfactory options that they can discuss.)

For example, where several social programs have impacts on jobs, health, public revenues, and political opinions, the alternative programs may have different components, such as cultural amenities, economic developments, and educational effects. How can such diverse, unrelated items be compared quantitatively?

Even if the costs and benefits of each item can be measured monetarily, there is still a problem. The subjective value of two actions with identical monetary outcomes may vary; health conditions may be perceived as more important than housing, housing more important than economics, and so on. One solution is to *weight* each issue or factor. Health conditions might be weighted as 2.0, housing conditions might be weighted 1.0, but economic conditions might be rated 0.5. Thus, the evaluation of any single course of action requires adding (3.5 x *Economic Value*) plus (2.0 x *Health Value*) plus (0.5 x

RANKED VALUES FOR ALTERNATIVE COMMUNITY PROJECTS

COMMUNITY PROJECT REQUESTS	HEALTH FACTORS (WEIGHT= 2.0)		HOUSING FACTORS (WEIGHT= 1.0)		ECONOMIC FACTORS (WEIGHT= 0.5)		TOTAL VALUE	RANK ORDER
	BASE VALUE # OF PEOPLE SERVED (IN 1000)	WEIGHTED VALUE	BASE VALUE # OF PEOPLE SERVED (IN 1000)	WEIGHTED VALUE	BASE VALUE # OF PEOPLE SERVED (IN 1000)	WEIGHTED VALUE		
NEW CLINIC	40	80	0	0	0	0	80	7
HOSPITAL EXPANSION	60	120	0	0	2	1	121	2
ELDERLY HOUSING	20	40	25	25	50	25	90	4
LOW INCOME HOUSING	30	60	30	30	12	6	96	3
NURSING HOMES	50	100	50	50	0	0	150	1
FOOD SUPPY PROGRAM	35	70	0	0	10	5	85	6
JOB TRAINING PROGRAM	0	0	10	10	80	40	50	8
WELFARE PAYMENT SUPPLEMENT	10	20	60	60	18	9	89	5

FIGURE 5.27 SIMPLE WEIGHTS AND RANKS (*Explanation:* The basic technique for combining and comparing noncomparable values is to assign subjective weights to each factor, multiply the weight by an associated value, sum the weighted values, and convert these sums to a rank order. The final ranks can then be used as the values for a decision matrix or hierarchy.

Housing Value). Of course, this approach assumes that all the three components—health, housing, economics—are measured in comparable terms such as dollars. They could also be measured, comparably, in terms of the numbers of people served (Figure 5.27).

There are many ways to integrate ranking and weighting techniques into the use of benefit-cost analyses, decision matrices, and decision hierarchies. There are also permutations of each technique. Given the complex nature of such presentations, it often is preferable to present independently the ranking and weighting of actions, outcomes, and similar items. Once the audience comprehends the procedure, the ranks or weights can be folded into the other decision-making frameworks.

Sensitivity to Weight Fluctuations

Portraying the sensitivity of the analysis to minor shifts in rankings or weights presents a standard problem. In the previous example, if economic conditions were weighted 0.7 instead of 0.5, would this change the final conclusions, and what would happen if they were rated 1.0, 0.8, or 0.6? The issue here is that many audiences are aware of the subjective, sometimes arbitrary, attachment of numerical values to ordinal rankings. The audience might easily comprehend the argument that economic conditions are less important than the others, but will they comprehend why they are weighted precisely 0.5?

There are procedures for deriving relatively precise weights by using a lengthy series of paired comparisons between sets of items (see Appendix for references). Even when such procedures are used, however, there are still minor, unexplained fluctuations in weights. In presenting such an analysis, it may be useful to show the effect of these fluctuations to the audience by using a weight range rather than a single number, computing the values and indicating where, if at all, these fluctuations affect the final analysis (Figure 5.28). In practice, these fluctuations may change the numbers, but have little effect on the rationality of picking one item over another. If they have such an impact, it is important to demonstrate this to an audience. For example, the analysis may show that two program options are superior to all the others, but it may not demonstrate that one of these two superior options is the absolute best. Such a conclusion does not beg the question, but simply indicates that the weighting methodology is an inappropriate technique to make a final choice, being appropriate only to narrow down the range of satisfactory or preferred actions. Therefore, the technique should not be pushed beyond its limits.

The sensitivity of an analysis to fluctuations in weights may also be shown by varying one particular weight that is relevant to the audience, and showing how it influences the final conclusion. It may be effective to show the upper and lower limits at which fluctuations in the weight change the conclusions. For example, it may be that changing the weight of health conditions from 2.0 to 1.5 or 2.5 in the situation previously described does not effect the final conclusions (Figure 5.29).

TABLE A.   ANALYSIS AND RECOMMENDATION FOR INVESTMENT OPTIONS

INVESTMENT OPTIONS	FINAL RANK (5 = BEST)	TOTAL WEIGHTED VALUE	EXPECTED PROFIT WEIGHT = 6			TIME NEEDED FOR COMPLETION WEIGHT = 3			POLITICAL DIFFICULTY WEIGHT = 2			TECHNICAL DIFFICULTY WEIGHT = 2		
			WEIGHTED VALUE	RANK	BASE EVALUATION ($1,000)	WEIGHTED VALUE	RANK	BASE EVALUATION	WEIGHTED VALUE	RANK	BASE EVALUATION	WEIGHTED VALUE	RANK	BASE EVALUATION
A	1	37.0	21	3.5	500	9.0	3.0	18 months	3	1.5	LOW	4	2.0	LOW
B	2	37.5	6	1.0	2,000	13.5	4.5	24 months	9	4.5	HIGH	9	4.5	HIGH
C	3	38.5	21	3.5	500	4.5	1.5	12 months	9	4.5	HIGH	4	2.0	LOW
D	4	40.5	12	2.0	1,000	13.5	4.5	24 months	6	3.0	MOD-ERATE	9	4.5	HIGH
E	5	41.5	30	5.0	400	4.5	1.5	12 months	3	1.5	LOW	4	2.0	LOW

TABLE B.   ALTERNATIVE ANALYSIS

COMPARISON OF RANKS				WEIGHTS (BASED ON CONSERVATIVE FORECAST)				
INVESTMENT OPTIONS	MAJOR DIFFERENCES	RANK (PROPOSED)	RANK (CONSERVATIVE)	WEIGHTED VALUE	EXPECTED PROFIT WEIGHT = 5 NEW WEIGHTED VALUE *	TIME NEEDED FOR COMPLETION WEIGHT = 4 NEW WEIGHTED VALUE *	POLITICAL DIFFICULTY WEIGHT = 3 NEW WEIGHTED VALUE *	TECHNICAL DIFFICULTY WEIGHT = 3 NEW WEIGHTED VALUE *
A		1	1	40.0	17.5	12	4.5	6.0
B	√	2	4	50.0	5.0	18	13.5	13.5
C		3	3	43.0	17.5	6	13.5	6.0
D		4	5	50.5	10.0	18	9.0	13.5
E	√	5	2	41.5	25.0	6	4.5	6.0

* COMPUTED USING SAME RANKS AND
BASE EVALUATIONS AS PRIOR ANALYSIS

FIGURE 5.28   COMBINING NONCOMPARABLE VALUES   (*Explanation:* In many situations, the factors to be combined are measured in different values. In these cases the values can first be converted to ranks, then multiplied by the weight, totaled, and given a final combined rank. It is also useful to show the audience the impact of variations in the weights; that is, the *sensitivity* of the final outcomes to changes in weights. These two tables offer one way to make these comparisons.)

Where many outcomes, programs, objectives, or actions are being ranked and weighted, the only important cases will be the marginal ones. Minor changes in weights or weighting procedures may not influence the ranking of items at the top or bottom of a list of objectives or program outcomes, but they may affect the few cases toward the middle of the list. This can be critical. If the budget for a program can include only the 10 highest ranked actions out of a list of 20, actions near the top are clearly included, and those at the bottom excluded. However, for those in the middle, minor changes in computational procedure have dramatic consequences. In these situations, showing the sensitivity of fluctuation in weights should focus on the marginal cases (Figure 5.30).

SENSITIVITY OF FUNDING RECOMMENDATIONS TO HEALTH FACTORS

COMMUNITY PROJECT REQUESTS	RECOMMENDED FINDING PRIORITIES	FINAL RANKINGS			TOTAL WEIGHTED VALUE			HEALTH FACTORS				HOUSING FACTORS		ECONOMIC FACTORS	
		HIGH HEALTH WEIGHT	MODERATE HEALTH WEIGHT	LOW HEALTH WEIGHT	WITH HIGH WEIGHT FOR HEALTH FACTOR	WITH MD WEIGHT FOR HEALTH FACTOR	WITH LOW WEIGHT FOR HEALTH FACTOR	WEIGHTED VALUE (high) (2.5)	(m.) (2.0)	(low) (1.5)	BASE VALUE # of people served	W.V. (1.0)	B.V. # of people served	W.V. (0.5)	B.V. # of people served
NURSING HOMES		1	1	1	175	150	125	125	100	75	50	50	50	0	0
HOSPITAL EXPANSION		2	2	2	151	121	91	150	120	90	60	0	0	1	2
LOW INCOME HOUSING	√	3	3	4	111	96	81	75	60	45	30	30	30	6	12
ELDERLY HOUSING	√	5.5	4	5	100	90	80	50	40	30	20	25	25	25	50
WELFARE PAYMENT SUPPLEMENT	√√	7	5	3	94	89	84	25	20	15	10	60	60	9	18
FOOD SUPPLY PROGRAM	√√	4	6	6	102.5	85	67.5	87.5	70	52.5	35	0	0	15	30
NEW CLINIC	√	5.5	7	7	100	80	60	100	80	60	40	0	0	0	0
JOB TRAINING PROGRAM		8	8	8	30	30	30	0	0	0	0	10	10	20	40

√√ THESE PROJECTS PRIORITIES ARE ESPECIALLY SENSITIVE TO HEALTH FACTORS

√ THESE PRIORITIES ARE SOMEWHAT SENSITIVE TO HEALTH FACTORS

WV = WEIGHTED VALUE

BV = BASE VALUE

FIGURE 5.29 SENSITIVITY OF WEIGHTS (**Explanation:** Audience decision makers are frequently the people who subjectively set the weights for each factor. They should understand the consequences of these judgments. The following table illustrates how an analyst might present the sensitivity of a recommendation to fluctuations in one set of weights.)

These problems also may be resolved by using the weights and numeric values only at the beginning of the analysis. The presentor should emphasize that the computational procedures are no more than a heuristic device, or rule of thumb, for placing items in relative order. Once the quantitative order is established, a final ranking includes only ordinal labels such as high, moderate, and low (Figure 5.31), and the original quantitative measures can be discarded. The analyst and/or audience reviews the items in each ordinal category, although the possibility of moving actions from one category to another should be considered. The argument for doing this is based on the grounds that the initial computational procedures were not inclusive of all factors, and that there is always some unique circumstance that may be critical.

FUNDING RECOMMENDATIONS (PROPOSED VS. PREVIOUS EVALUATION SYSTEM FOR TRANSPORTATION PROJECTS)

PROJECT KEY	COST (IN $1,000) FUNDING CUT-OFF 20,000,000		SENSITIVITY ANALYSIS				SUPPORTING DATA					
			PROPOSED WEIGHTING SYSTEM SAFETY=2.4 ECONOMICS=1.9 SERVICE=1.5		PREVIOUS WEIGHTING SYSTEM SAFETY=3.0 ECONOMICS=0.5 SERVICE=2.1		UNWEIGHTED RANK (BEST=20)	SAFETY FACTOR RATING (MAX 100 PTS)	UNWEIGHTED RANK (BEST=20)	ECONOMIC BENEFITS (IN $1,000)	UNWEIGHTED RANK (BEST=20)	SERVICE FACTOR # OF PEOPLE W/ IMPROVED SERVICE (IN 1,000)
			RANK BEST=20	WEIGHTED VALUE	RANK BEST=20	WEIGHTED VALUE						
A	3800		20	101.3	20	92.5	17	90	20	7650	15	19.8
B	900		19	93.0	19	86.1	14.5	85	18	6500	16	20.0
C	850		18	78.6	17	80.5	9	72	15	5950	19	80.5
D	1900	X	17	75.7	11	63.5	4	59	19	6750	20	85.0
E	5800		16	74.9	16	69.4	14.5	85	14	5900	9	9.5
F	2150		15	74.2	13	66.8	7	70	16	6000	18	70.2
G	1500		14	73.6	18	74.6	19	91	10	3750	6	6.5
H	930	X	13	72.2	10	62.2	6	69	17	6100	17	50.0
I	1000		12	71.1	15	69.3	12	80	12	3900	13	18.3
J	980		11	69.2	14	67.0	17	90	11	3800	5	5.0
K	670		10	58.4	9	60.7	10.5	78	8	3650	12	17.3
L	1750	✓	9	57.1	12	64.1	20	92	4	900	1	3.0
M	500		8	49.1	8	58.3	17	90	2	200	3	4.5
N	1050		7	43.7	7	47.5	8	71	5	1900	10	10.0
O	1200		6	42.9	5	39.9	2	55	9	3700	14	19.2
P	2000		5	42.4	2	30.2	3	56	13	4000	7	8.5
Q	900		4	37.6	6	45.8	13	83	1	100	3	4.5
R	800		3	37.3	3	35.3	5	63	7	3500	8	9.0
S	700		2	35.4	4	39.3	10.5	78	3	500	3	4.5
T	600		1	30.3	1	29.1	1	51	6	2000	11	16.0

(The column between COST and SENSITIVITY ANALYSIS is labelled RECOMMENDED FOR FUNDING for the upper rows (A–L) and NOT RECOMMENDED FOR FUNDING for the lower rows.)

KEY: **X** = WOULD NOT HAVE BEEN FUNDED UNDER PREVIOUS WEIGHTING SYSTEM

✓ = WOULD HAVE BEEN FUNDED UNDER PREVIOUS WEIGHTING SYSTEM

FIGURE 5.30 IMPACT OF WEIGHTS (**Explanation:** The most significant aspect of a sensitivity analysis for the audience is the final impact on the decision-making process. In this illustration the impact of the two weighting systems is presented in terms of projects that would or would not be funded, given each set of weights.)

In some circumstances, the weighting procedure might indicate major gaps in relative values, so that some items are rated *very high*, some *high*, and some *low*, with no items rated *moderate*. This indicates to an audience that, although the ranking is only ordinal, there are implied differences in the jump from one level to the next (Figure 5.31).

SUBJECTIVITY OF WEIGHTS

The weighting of outcomes and objectives will vary with the person or group making the subjective judgments. There is nothing wrong with this, as it is an inherent, positive aspect of the procedure. Nevertheless, the presentor should

TABLE A   FUNDING PRIORITIES FOR COMMUNITY PROJECTS

COMMUNITY PROJECT REQUESTS	FINAL EVALUATION AND PRIORITY	HEALTH FACTORS WEIGHTED VALUE			HOUSING FACTORS WEIGHTED VALUE	ECONOMIC FACTORS WEIGHTED VALUE
		METHOD A	METHOD B	METHOD C		
NURSING HOMES	HIGH	125	100	75	50	0
HOSPITAL EXPANSION	HIGH	150	120	90	0	1
LOW INCOME HOUSING	MODERATE	75	60	45	30	6
ELDERLY HOUSING	MODERATE	50	40	30	25	25
WELFARE PAYMENT SUPPLEMENT	MODERATE	25	20	15	60	9
FOOD SUPPLY PROGRAM	MODERATE	87.5	70	52.5	0	15
NEW CLINIC	LOW	100	80	60	0	0
JOB TRAINING PROGRAM	LOW	0	0	0	10	20

FIGURE 5.31 PRESENTING ORDINAL CATEGORIES (*Explanation:* Ranking and weighting systems may seem too abstract for some audiences. In these situations the final numbers can be converted to ordinal categories such as high, medium, and low. These two Tables are based on Figures 5.29 and 5.30. In both cases the data have been summarized and less data are shown. In Table B the ranks have been renumbered (1 is the highest rank instead of 20) to make the presentation more sensible. Also, in Table B, the ordinal labels help clarify the big gap in values between projects rated "high" versus those rated "low.")

be prepared to display how such subjectivity effects the analysis, which is a good way to portray different sets of audience interests. For example, one set of weights, and the associated conclusion, may represent one set or coalition of audience interests. A second set of weights represents another interest group, and so on. The analyst can then show the decision that would result from combining and/or averaging the weights representing different positions, opinions or interests (Figure 5.32).

Doing this also can lead to the identification of those weights that are most controversial, identifying where there is the greatest variance in the weights representing different subjective viewpoints. If considerable variation in the weights does not influence the final outcome, it can help avoid time consuming debate and discussion. Alternatively, if significant variation

TABLE B   FUNDING PRIORITIES FOR TRANSPORTATION PROJECTS

FINAL EVALUATION			PRIORITY RANKING		
PROJECT KEY	FUNDING PRIORITY *	WEIGHTED SCORE	SAFETY FACTORS (WEIGHT=2.4) RANK=	ECONOMIC FACTORS (WEIGHT=1.9) RANK=	SERVICE FACTORS (WEIGHT=1.5) RANK=
A	VERY HIGH	101.3	4	1	6
B	VERY HIGH	93.0	6.5	3	5
C	HIGH	78.6	12	6	2
D	HIGH	75.7	17	2	1
E	HIGH	74.9	6.5	7	12
F	HIGH	74.2	14	5	3
G	HIGH	73.6	2	11	15
H	HIGH	72.2	15	4	4
I	HIGH	71.1	9	9	8
J	HIGH	69.2	4	10	16
K	LOW	58.4	10.5	13	9
L	LOW	57.1	1	17	20
M	LOW	49.1	4	19	18
N	LOW	43.7	13	16	11
O	LOW	42.9	19	12	7
P	LOW	42.4	18	8	14
Q	LOW	37.6	8	20	18
R	LOW	37.3	16	14	13
S	LOW	35.4	10.5	18	18
T	LOW	30.3	20	15	10

(FUNDED: rows A–J; NOT FUNDED: rows K–T)

* VERY HIGH = SCORE ABOVE 79.0
  HIGH = 69.0 TO 79.0
  MODERATE = 59.0 TO 69.0
  LOW = BELOW 59.0

TABLE A   FUNDING PRIORITIES FOR COMMUNITY PROJECTS

COMMUNITY PROJECT REQUESTS	POTENTIAL CONTROVERSY	FINAL PRIORITIES BY GROUP		
		CENTRAL ADMINISTRATION	NEIGHBORHOOD GROUPS	GENERAL PUBLIC (TELEPHONE SURVEY)
NURSING HOMES		HIGH	MODERATE	MODERATE
HOSPITAL EXPANSION	✱	HIGH	LOW	MODERATE
LOW INCOME HOUSING	✱	MODERATE	HIGH	LOW
ELDERLY HOUSING		MODERATE	HIGH	MODERATE
WELFARE PAYMENT SUPPLEMENT		MODERATE	MODERATE	LOW
FOOD SUPPLY PROGRAM	✱	MODERATE	LOW	HIGH
NEW CLINIC		LOW	LOW	LOW
JOB TRAINING PROGRAM	✱	LOW	HIGH	MODERATE

RELATED ANALYSIS					
CENTRAL ADMINISTRATION			NEIGHBORHOOD GROUPS		GENERAL PUBLIC
HEALTH FACTORS (RANK ORDER)	HOUSING FACTORS (RANK ORDER)	ECONOMIC FACTORS (RANK ORDER)	HEALTH FACTORS (RANK ORDER)	SOCIO-ECONOMIC RATING	% IN FAVOR IN TELEPHONE SURVEY
2	2	7.5	6	MODERATE	33%
1	7	6	7	MODERATE	25%
5	3	5	1	HIGH	15%
6	4	1	2	MODERATE	40%
7	1	4	4	MODERATE	20%
4	7	3	5	LOW	65%
3	7	7.5	8	LOW	5%
8	5	2	3	HIGH	48%

TABLE B   INVESTMENT RECOMMENDATIONS BASED ON CONSENSUS

SUMMARY OF EVALUATION BY GROUPS					POTENTIAL INVESTORS					LAND OWNERS			PUBLIC OFFICIALS				PLANNING CONSULTANT		
INVESTMENT OPTIONS	POTENTIAL INVESTORS	LAND OWNER	PUBLIC OFFICIALS	PLANNING CONSULTANTS	COMBINED RANKS	EXPECTED PROFIT	TIME FOR COMPLETION	POLITICAL DIFFICULTIES	TECHNICAL DIFFICULTIES	COMBINED RANKS	EXPECTED PROFIT	COMPETING BIDS	COMBINED RANKS	POTENTIAL CONTROVERSY	EXPECTED REVENUE	ADMINISTRATIVE DIFFICULTY	COMBINED RANKS	LONG TERM VALUE	TECHNICAL DIFFICULTY
A	1	2	1	4.5	1	3.5	3	1.5	2	2	2	2	1	1.5	3	3	4.5	3.5	4.5
√ B	2	1	2.5	1	2	1	4.5	4.5	4.5	1	1	3	2.5	4.5	2	1	1	2	1
√ C	3	3	4	4.5	3	3.5	1.5	4.5	2	3	3	1	4	4.5	1	4	4.5	3.5	4.5
D	4	4	5	2	4	2	4.5	3	4.5	4	4	4	5	3	4.5	2	2	1	3
E	5	5	2.5	3	5	5	1.5	1.5	2	5	5	5	2.5	1.5	4.5	5	3	5	2

√ = RECOMMENDED PROJECTS: HIGH DEGREE OF CONSENSUS

FIGURE 5.32 RANKING BY GROUPS (*Explanation:* Audiences frequently contain decision makers with divergent values and objectives. In these circumstances, tables can be constructed to represent different value systems and how they may combine into a potential compromise. Table A builds on the illustrations in Figures 5.27, 5.29, and 5.31. Table B is based on Figure 5.28. In both examples previously presented data have been summarized so new information can be presented prominently.)

TABLE A   FUNDING PRIORITIES FOR TRANSPORTATION PROJECTS

PROJECT KEY		COST (IN $1,000)	FINAL RANK	TOTAL WEIGHTED SCORE	DECISION FACTORS		
					SAFETY FACTORS (WEIGHT=2.4) RATING*	ECONOMIC FACTORS (WEIGHT=1.9) RATING*	SERVICE FACTORS (WEIGHT=1.5) RATING*
RECOMMENDED FOR FUNDING	A	3800	1	11.8	1	1	5
	F	2150	2	12.1	3	1	2
	C	850	3	12.5	3	2	1
	B	900	4	12.7	2	1	4
	D	1900	5	15.4	5	1	1
	G	1500	6.5	15.6	1	3	5
	J	980	6.5	15.6	1	3	5
	H	930	8	16.0	4	1	3
	E	5800	9	16.1	2	2	5
	I	1000	10	18.0	2	3	5
NOT RECOMMENDED FOR FUNDING	L	1750	11	19.4	1	5	5
	M	500	12	19.4	1	5	5
	K	670	13	20.4	3	3	5
	Q	900	14	21.8	2	5	5
	N	1050	15	22.3	3	4	5
	O	1200	16.5	25.2	5	3	5
	P	2000	16.5	25.2	5	3	5
	R	800	18	22.8	4	3	5
	S	700	19	24.2	3	5	5
	T	600	20	27.1	5	4	5

* KEY FOR RATINGS

RATING	SAFETY BASED ON MAX POINT SCORE=100	ECONOMIC BASED ON BENEFITS (IN $1,000)	SERVICE BASED ON # OF PEOPLE SERVED (IN 1,000)
1	90-100 PTS	OVER 6000	OVER 80
2	80- 89 PTS	4500 TO 5999	60 TO 79.9
3	70-79 PTS	3000 TO 4499	40 TO 59.9
4	60- 69 PTS	1500 TO 2999	20 TO 39.9
5	50-59 PTS	0 TO 1499	0 TO 19.9

FIGURE 5.33   RANKS WITH RATINGS AND STANDARDIZED SCORES   (*Explanation:* Often it is advisable to convert noncomparable quantities into similar units of measurement *before* computing weighted scores. Tables A and B are both based on the identical data shown in Figure 5.30. Table A converts the raw data in Figure 5.30 to ratings of 1 to 5 before computing weighted scores. In Table B the same data are standardized by converting them to percentages; that is, the raw numbers in each column of Figure 5.30 are summed, and then each number is divided by its column total. In both Tables A and B the project key letters correspond to the identical letters in Figure 5.30. An examination of the new, nonalphabetical order of the letters indicates the degree to which this alternative ranking and weighting method can alter the results.)

TABLE B   FUNDING RECOMMENDATION FOR TRANSPORTATION PROJECTS
(BASED ON STANDARDIZED VALUES)

PROJECT KEY		COST (IN $1,000)	FINAL RANK	TOTAL WEIGHTED SCORE	SAFETY FACTORS (WEIGHT=2.4) (100 PT TOTAL)	ECONOMIC FACTORS (WEIGHT=1.9) (100 PT TOTAL)	SERVICE FACTORS (WEIGHT=1.5) (100 PT TOTAL)
RECOMMENDED FOR FUNDING	D	1900	1	5.37	3.9	8.8	18.4
	C	850	2	5.24	4.8	7.8	17.5
	F	2150	3	4.88	4.6	7.8	15.2
	H	930	4	4.23	4.6	7.9	10.8
	A	3800	5	3.97	6.0	10.0	4.3
	B	900	6	3.61	5.6	8.5	4.3
	E	5800	7	3.12	5.6	7.7	2.1
	I	1000	8	2.83	5.3	5.1	4.0
	K	670	9	2.71	5.2	4.8	3.8
	G	1500	10	2.59	6.0	4.9	1.4
NOT RECOMMENDED FOR FUNDING	J	980	11	2.54	6.0	5.0	1.1
	O	1200	12	2.42	3.6	4.8	4.2
	P	2000	13.5	2.16	3.7	5.2	1.8
	R	800	13.5	2.16	4.2	4.6	2.0
	N	1050	15	1.93	4.7	2.5	2.2
	T	600	16	1.83	3.4	2.6	3.5
	L	1750	17	1.78	6.1	1.2	0.7
	M	500	18	1.63	6.0	0.3	1.0
	S	700	19	1.51	5.2	0.7	1.0
	Q	900	20	1.49	5.5	0.1	1.0

* KEY FOR RATINGS

RATING	SAFETY BASED ON MAX POINT SCORE=100	ECONOMIC BASED ON BENEFITS (IN $1,000)	SERVICE BASED ON # OF PEOPLE SERVED (IN 1,000)
1	90-100 PTS	OVER 6000	OVER 80
2	80- 89 PTS	4500 TO 5999	60 TO 79.9
3	70-79 PTS	3000 TO 4499	40 TO 59.9
4	60- 69 PTS	1500 TO 2999	20 TO 39.9
5	50-59 PTS	0 TO 1499	0 TO 19.9

in the weighting of an item has major influence on the analytic results, this variation becomes an appropriate topic for dialogue and negotiation.

Integers, Percentages, Scales

In simpler situations, it is sufficient to use whole integers in assigning numeric weights. For example, outcome A is weighted 3, outcome B is weighted 2, outcome C is weighted 6, and so on, which is relatively easy to comprehend.

In situations where there are perhaps 10 or more items to be weighted, other approaches may be easier to present. One such approach is to use a scale where several items may be weighted 5 (very high), 4 (high), 3 (moderate), 2 (low), and 1 (very low). Five-, seven-, and ten-point scales are commonly used in this way (Figure 5.33).

It is sometimes forgotten that the higher number should be assigned to the most positive value. This avoids the minor ambiguity of explaining to the audience that the action or objective with the lowest number is really the highest value.

Another approach uses percentages where the total of all the weights equals 100 percent or 1.0, which can be explained as awarding a total of 100 points to all the actions or outcomes. The analysis may begin by assigning integers to each outcome or action, then adding up all the integers and dividing each integer by the total to arrive at a percentage (Figure 5.33). For instance, six outcomes weighted 8, 12, 4, 4, 10, 2 would be recomputed as 20%, 30%, 10%, 10%, 25%, 5% (or 0.2, 0.3, 0.1, 0.1, 0.25, 0.05). In some cases, different numerical systems may lead to different conclusions, depending on the arithmetic computations involved. Even when two different numerical systems lead to the same conclusion, one may be easier to comprehend and therefore create a more effective presentation.

6

Schedules, Budgets and Outcomes: Statistics for Project Implementation

Major projects and programs often entail complex relationships among time, personnel, funds, types of tasks, and other resources or constraints, as well as intended outcomes and objectives. Statistics are frequently used in presenting these issues. Although the techniques discussed in this chapter are not oriented toward programs that provide ongoing services or continuing activities, they are useful for managing projects that have a clear beginning and end.

Operations research, management science, organizational theory, and related disciplines have developed powerful quantitative analytic procedures for such situations, but these procedures can be difficult for audiences to comprehend and appreciate. A typical situation may involve a presentation to a committee that reviews the data to make a decision, such as whether to approve a project, modify a budget, organize a work team or hire a consultant.

Audiences reviewing these presentations may be familiar with some advanced statistical methods, although it is more likely that only a few members of the target audience are knowledgeable about the procedures involved. There also may be one or two advanced techniques with which no one is familiar. If so, the audience is unlikely to sympathize with their use or may be unable to comprehend their specific application, thus hindering the decision-making process further.

These typical situations suggest that presentations will be more effective if only simple techniques are used, or if sophisticated techniques are modified to be more direct, and less abstract and complicated. The following subsections illustrate some of the more useful techniques available to the analyst in developing project implementation statistics.

BAR DIAGRAMS: TIME, COST, PERSONNEL, OUTCOMES

The most widely used form of presentation is a type of bar diagram and table, where each bar, usually displayed horizontally, represents a specific task listed along the left-hand side. Successive time periods are illustrated across the top, where each bar has a start time and end time. Bar diagrams can integrate data on time, budget, personnel, outcomes, and the logical interrelations among these factors. Theoretically, any project can be differentiated into a minute-by-minute, dollar-by-dollar accounting of activities. It appears simple, but the complexity, amount, and type of information can create presentation difficulties, and the level of generalization or differentiation becomes an important issue. The presentor should only display the relevant information without insignificant detail, especially to a small audience that has some familiarity with the issues in question (Figure 6.1).

Time Display

In its simplest version, a bar diagram shows only the amount of time within which a task will be completed. Usually, however, there are several logical

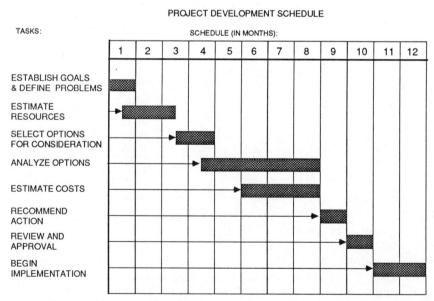

FIGURE 6.1   SIMPLE TIME BAR   (**Explanation:** Often, simple time bar diagrams, without extraneous detail, are an effective way to communicate a project management plan to an audience with prior knowledge of the type of problems and issues being considered.)

relationships between tasks that also need to be presented; the completion of one task may be necessary before another task can begin, or if one task leads into three others. In this case, three lines showing that relationship will be sufficient. The converse situation also may occur where three tasks lead into one subsequent activity (Figure 6.2). If the relationships are extremely complex, a network diagram, as described in the next section, may be more appropriate.

When a whole set of tasks precedes another set, it may be useful to show the project in phases, labeled across the top. The end of the first set of tasks is the end of Phase I and the beginning of Phase II. To make the presentation more effective, it may be useful to name *and* number each phase, such as *preliminary planning, feasibility analysis,* or *implementations* (Figure 6.3).

In addition to different phases, there may be important events that are repeated. For example, there may be several times when activities require formal review and approval. These can be shown with a special symbol at the end of the tasks where such events occur. Another option may be to show such key events with lines dividing one phase or set of tasks from all subsequent tasks.

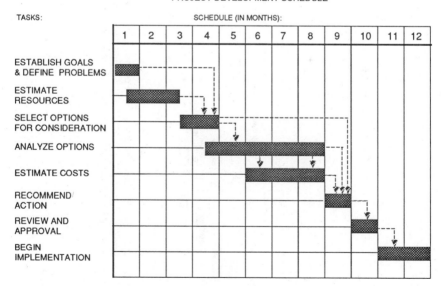

FIGURE 6.2   TIME LINES AND TASK RELATIONSHIPS   **(*Explanation:*** A common addition to most time-line diagrams is an indication of the logical relationship between tasks. Often these relationships are general, indicating only that one task "feeds into" another. This diagram is an elaboration of Figure 6.1.

PHASED PROJECT DEVELOPMENT SCHEDULE

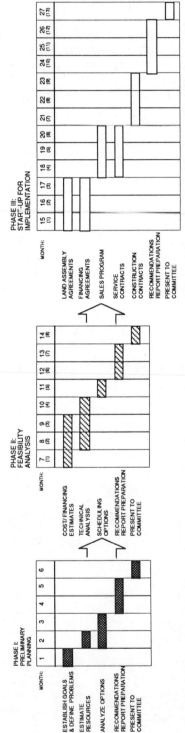

FIGURE 6.3 SEGMENTED SCHEDULES (*Explanation:* In some projects it may be more effective to segment a management plan or schedule into distinct phases. This is especially useful in longer projects or in projects where there are relatively clear separations among the sets of tasks because of the personnel who are assigned to the task, the need for formal decision-making reviews, or other organizational constraints.)

Similarly, there may be external events or activities beyond the control of the organization undertaking the tasks that are essential to the continuation of the work program. Here, too, showing a major break between sets of tasks can be useful in presenting the uncertainty created by outside occurrences (Figure 6.4).

The time between the beginning and end of a task is not necessarily equivalent to the time required to perform it. If a task requiring three weeks of work begins in week 1, but need not be completed until the end of week 7, the implication arises that during the seven-week period, the people involved in that task have some *slack* (or time allocated to tasks for other projects, programs, and organizational activities).

Some tasks can be started, but not completed until other tasks are also completed. These kinds of situations can be presented by using a dashed bar to indicate that the performance of the task is not continuous (Figure 6.5). In addition, lines drawn from other tasks can be used to portray logical relationships. If there are too many discontinuities, it may be more effective to subdivide some tasks into two or more subtasks, each with its own visually discrete bar in the diagram.

In addition to discontinuities in tasks, there may be uncertainties in time allocations. The analyst may estimate the length of one task with considerable certainty based on a long track record of similar tasks. For other tasks, however, there may be a significant uncertainty because of the unfamiliarity of the task, new personnel, or new procedures. Here, small amounts of slack can be built into estimates without special emphasis. Uncertainties may also exist if there is a major difference in the upper and lower limits of the time required to perform a task; where one component of the project takes anywhere from two to twelve weeks, the difference may be as much as 600 percent. Such gross uncertainties should not be hidden if they are legitimate, but should be displayed clearly in the presentation with dashed lines, symbols, or codes (Figure 6.6).

Inclusion of Costs

Project management presentations frequently need to show costs figures in relation to time periods. Inclusion of cost figures on each bar is fairly straightforward, as the figures can be shown at the beginning, end, on top of, or within the bar, depending on the graphic technique being used. It often is useful to include summary costs figures or subtotals, and the subtotal showing costs for each phase may also be included.

In addition, some tasks may have uncertain costs or cost ranges that should be displayed. This is usually the case when the time for the task is itself uncertain, or in situations where costs, but not time, are dependent on outside factors such as inflation in the cost of materials and fees (Figure 6.7).

Another important issue to be considered in presenting cost figures is the potential for trade-offs between time and money. For example, a task may take

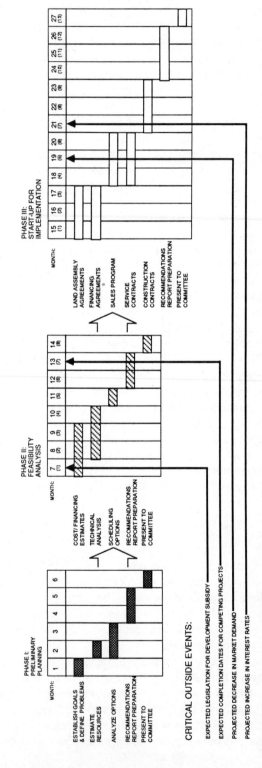

PHASED PROJECT DEVELOPMENT SCHEDULE AND KEY EVENTS

FIGURE 6.4 SCHEDULE WITH OUTSIDE EVENTS (*Explanation:* In some cases there are critical outside events that should be viewed simultaneously with a project schedule. In this diagram, based on Figure 6.3, additional events can be reviewed to see how they might effect the proposed project plan.)

182

DIAGRAM A: DETAILED SCHEDULE

PHASE II: Feasibility Analysis

DIAGRAM B: DETAILED SCHEDULE

PHASE II: Feasibility Analysis

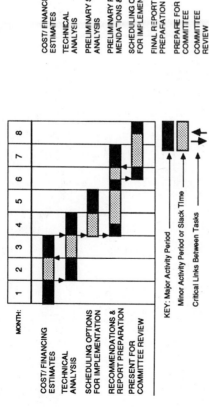

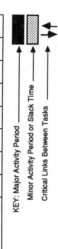

FIGURE 6.5 DIFFERENTIATING LEVELS OF ACTIVITY (*Explanation:* Using different types of line gradations can help distinguish different levels of activity. Both of these diagrams are elaborations of one part of the time-line diagram shown in Figure 6.3. Diagram B, however, actually splits apart some of the original activities into completely distinct tasks. The degree to which activities are linked, separated, or shown at different levels should depend on the complexity of the situation being described and the sophistication of the audience.)

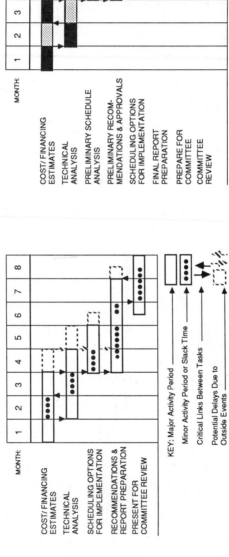

FIGURE 6.6 DISPLAYING UNCERTAINTY (*Explanation:* These two diagrams illustrate graphic techniques to display the uncertainty of the completion date for specific tasks. Both diagrams correspond to those of Figure 6.5.)

184

four weeks and cost $2000, but it may be possible to complete the work in one week, but at a total cost of $4000. This is typical of many projects. In some circumstances, therefore, there will be two or more estimates of the cost and completion time for each task, where each combination of completion times and budgets implies a unique project schedule. Thus, if each of 10 tasks has two time and cost estimates, there are over 1000 combinations for project work programs.

There are techniques for analyzing such situations and producing estimated completion times and budgets (noted in the next section). In practice, however, there may be only a few critical tasks for which time/money trade-offs are likely to be of serious interest to the audience. For instance, if the project completion time may be perceived by some decision makers as too long, the diagram could identify those tasks that are likely to be the best targets for reducing the total completion time at a reasonable cost (Figure 6.8). Often, visual inspection of a bar diagram combined with experienced subjective judgment is an effective presentation device.

Time/money trade-offs can be presented in two separate diagrams. One can show the minimum cost and the longer time, while the other shows minimum time at the higher cost. In simple situations, it may be sufficient to show everything on one diagram, using two adjacent bars for the targeted tasks. One bar represents the recommended or assumed time/cost combination and the other displays the shorter *crash* combinations of a shorter time and higher cost (Figure 6.9).

Inclusion of Personnel

In addition to time and money, the issue of human resources must be considered. Work programs often identify the type and/or number of personnel assigned to a task, which is particularly useful when projects involve different teams, organizational subunits, or even entirely different organizations collaborating on a project. The audience frequently wants to know *who* is going to do something in addition to when it will occur and how much it will cost.

The simplest case is one in which each task can be assigned to one team, personnel unit, or individual. This can be noted at the beginning, the end, or along the bar representing the task (Figure 6.10). The information can also be combined with costs, a technique that can accommodate a limited amount of information, such as a few tasks involving more than one type of personnel or team. In some cases, entire sets of tasks or phases can be identified with separate groups of people, and different nodes or symbols can be used to identify the personnel assigned to each task.

Other situations are more complex. there may be different combinations of major personnel who are assigned to multiple sets of tasks. Similarly, there may be different cost figures tied to different types of personnel that are critical

DIAGRAM A: DETAILED SCHEDULE AND COSTS

PHASE II: Feasibility Analysis

DIAGRAM B: DETAILED SCHEDULE AND COSTS

PHASE II: Feasibility Analysis

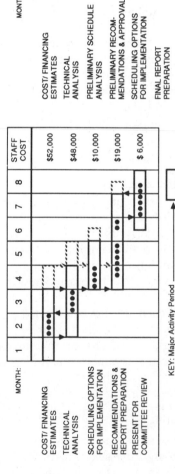

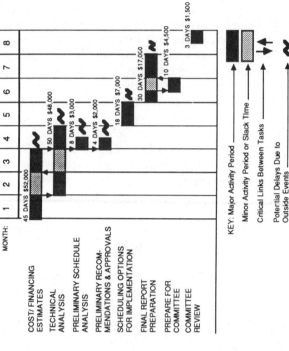

FIGURE 6.7 ADDING COSTS TO TIME DIAGRAMS (*Explanation:* The simplest way to portray costs is to attach them to each individual task. These two options build on the information shown in Figure 6.6.)

items of information. In these cases, it may be more effective to see several tables or diagrams. For example, one bar diagram could display the task completion times, cost figures, and principal personnel. A separate table could then display the same tasks vertically along one side, and list all personnel along the other dimension. The cells would contain the amount of time that each personnel category is devoting to the corresponding task and the associated cost. In addition, subtotals could be shown summarizing data for sets of tasks and/or sets of personnel.

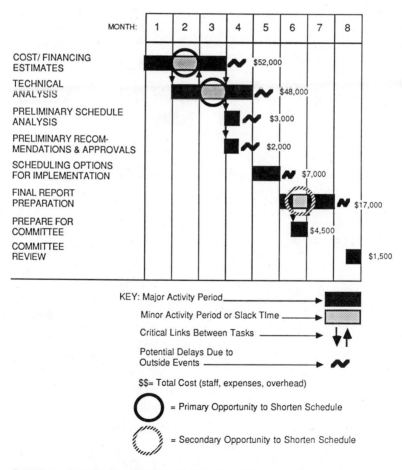

DIAGRAM B: DETAILED SCHEDULE AND COSTS FOR

PHASE II: Feasibility Analysis

KEY: Major Activity Period

Minor Activity Period or Slack TIme

Critical Links Between Tasks

Potential Delays Due to Outside Events

$$= Total Cost (staff, expenses, overhead)

◯ = Primary Opportunity to Shorten Schedule

◯ = Secondary Opportunity to Shorten Schedule

FIGURE 6.8 TARGETING CHANGES (*Explanation:* Scheduling diagrams are frequently used as tools for targeting tasks that should receive special attention. This is one example based on Figure 6.7.)

It also is possible to combine the table and bar diagram. The table is located on the left and the bar diagram on the right. The vertical list of tasks is situated in the middle, serving both diagrams. This helps focus the audience on the tasks as the key issue, and also helps them to integrate the data in a comprehensible way (Figure 6.11).

Inclusion of Outcomes

Any project or work program presumably has a set of objectives or desired outcomes. They all may occur at the end of the project and depend on

DIAGRAM A: SCHEDULE AND COSTS OPTIONS

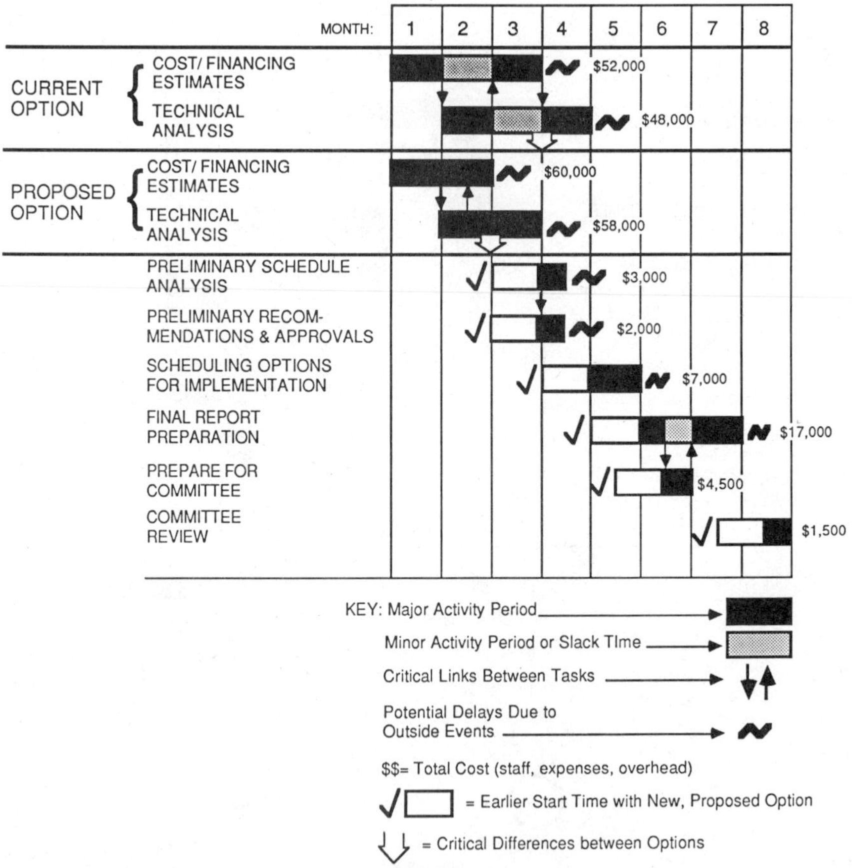

FIGURE 6.9 SHOWING SCHEDULING/COST OPTIONS (***Explanation:*** Simple options can be shown as an offshoot or component of a larger illustration (Diagram A). When options are more complex, they should be shown as two complete diagrams. In this situation, Diagram B would be used *in addition* to the diagram in Figure 6.8.)

the completion of all of the tasks. They can be shown with a statement or other visual cue at the end of the bar diagram (Figure 6.12), so that the audience can see the goals as the information is presented.

In other instances, different goals may be associated with the completion of only one task or subset of tasks. For example, a social program may have several components that are relatively independent of one another. The completion of one component represents, by itself, achievement of an objective or set of objectives. When this occurs, it may be more relevant to show the audience how different objectives are tied to different tasks *and* their associated completion times, costs, and personnel.

DIAGRAM B: PROPOSED SHORTEST SCHEDULE

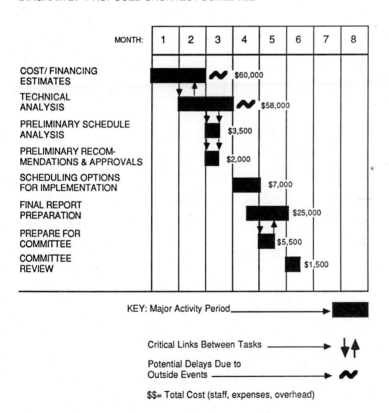

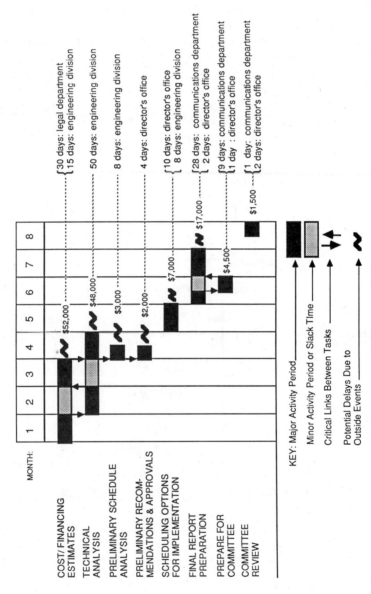

FIGURE 6.10 SIMPLE PERSONNEL ASSIGNMENTS (*Explanation:* Each time line can be linked visually to simple personnel assignments. This example elaborates on the information given in Figure 6.7.)

PROJECT MANAGEMENT FOR FEASIBILITY STUDY

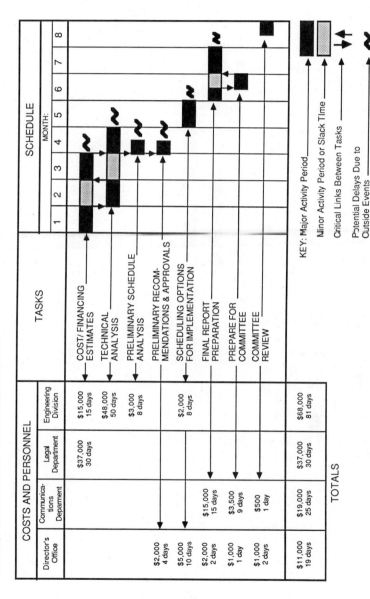

Figure 6.11 Cost, Personnel, and Schedules *(Explanation:* Time-line diagrams lend themselves to being combined with other types of diagrams. In this case the time-line schedule is combined with a table that displays personnel assignments and costs associated with each task, elaborating on the situation portrayed in Figure 6.10.)

DEVELOPMENT PROCESS FOR COMMUNITY FACILITY

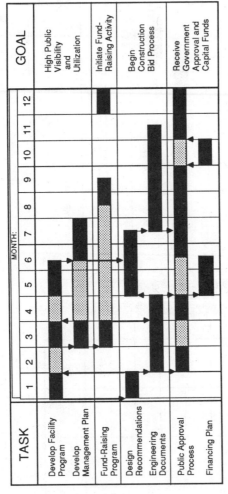

FIGURE 6.12 SCHEDULES AND GOALS/OUTCOMES (*Explanation:* In simple situations the goals or outcomes of individual activities or sets of activities can be displayed in a list at the end or side of the time-line diagram.)

DEVELOPMENT PROCESS FOR COMMUNITY FACILITY

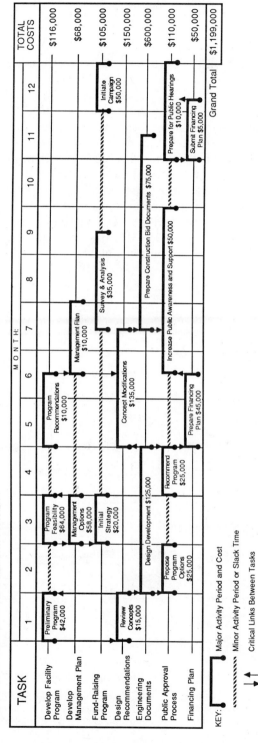

FIGURE 6.13 SCHEDULES AND SEGMENTED OUTCOMES (*Explanation:* In more complex situations, goals or outcomes can be written into each subtask or activity. This diagram shows such an approach by using the same situation that is shown in Figure 6.12.)

This can be effected with symbols or codes placed at the point in the diagram at which an objective is achieved, or by subdividing the diagram subsets of tasks (Figure 6.13). In other cases, objectives may be independent of each other, but may entail substantially the same tasks. For example, there may be one set of *preliminary planning, feasibility*, and *implementation* tasks that help achieve independent objectives. If the objectives are sufficiently independent, a separate diagram for each might be appropriate. Alternatively, one diagram could be used where the bar representing each task is subdivided to indicate the rough percentage of effort associated with each objective (Figure 6.14).

Another option utilizes a bar diagram with an accompanying table. The bar diagram would show the completion times for each task, and the table, which can be visually overlapped with the bar diagram, crosstabulates how the tasks relate to each objective (Figure 6.15). In some instances, it is sufficient to show only the table that crosstabulates objectives and tasks, and the bar diagram can be omitted.

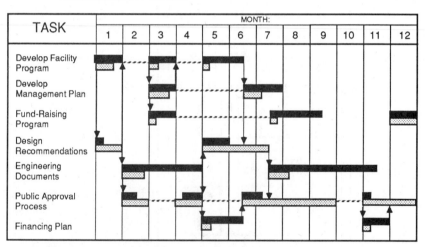

TASK SCHEDULE FOR FINANCING FEASIBILITY
AND COMMUNITY IMPACT

TASK	MONTH:											
	1	2	3	4	5	6	7	8	9	10	11	12
Develop Facility Program												
Develop Management Plan												
Fund-Raising Program												
Design Recommendations												
Engineering Documents												
Public Approval Process												
Financing Plan												

KEY:

▮ Major Activity Related to <u>Financial Feasibility</u>

▦ Major Activity Related to <u>Community Impact</u>

↓↑ Critical Links Between Tasks

⋯ Minor Activity

FIGURE 6.14 SPLITTING ACTIVITIES BY OBJECTIVES (*Explanation:* When there are only two or three key goals it may be effective to split each activity bar or line into component parts that show the rough proportion of time or effort related to each goal. This illustrates another version of the situation shown in Figure 6.12.)

A variation of this technique occurs when goals and objectives are structured hierarchically, like a decision tree. For example, the left most point in the diagram could be the principal, single goal. This would branch out into subgoals and each of these could branch out to lower-level objectives. At this point, the list of objectives can become one dimension of a table. Tasks are then the other (usually horizontal) dimension, where the cells contain information on costs, personnel, and so on (Figure 6.16).

NETWORKS: TIME, COST, PERSONNEL, OUTCOMES

Two of the best-known techniques for project management are CPM (Critical Path Method) and PERT (Program Evaluation and Review Technique). Both techniques are relatively sophisticated, requiring computational procedures beyond the comprehension of most general audiences. Application of these techniques, like other branches of statistical decision theory, requires conformance to logical constraints, assumptions, data needs, and so on. Computational procedures and related issues can be found in a number of existing texts (see Appendix for references).

Unlike other complex techniques, the results of PERT/CPM are not as difficult to present—it is the process of arriving at the results that is often obscure. Thus, if the data are available and all the logical constraints can be met, these techniques can be used and presented in a relatively unmodified way.

Typically, presentation consists of a network diagram in which there are several *nodes* (often shown as circles or squares) and *flows* (shown as lines from one node to another). Nodes are the events or moments in time when activities start and end. The flows have arrowheads to indicate their direction, and represent activities. Thus, a PERT/CPM diagram has a start point from which activities emanate, and one end point at which all flows converge, signaling final completion of the project. Each flow is labelled with the time it takes and the name (or code) of the activity involved. Nodes also may have specific names attached to them, but are usually just given a reference number. Similarly, flows or activities also may have reference numbers. The amount of information can become visually confusing (Figure 6.17), so all of the items should not be given equal emphasis. In most instances, the most important elements will be the name of the activity and the time required, which should appear as the most graphically dominant.

The presentation techniques described in the following sections are, in some cases, compatible with the formal use of CPM and PERT methodologies, although the presentation techniques may include information and/or logical relations that cannot be accommodated by traditional methods of network analysis. Such techniques are included here because they represent effective ways of communicating information, even though they do not allow for more rigorous quantitative methodologies.

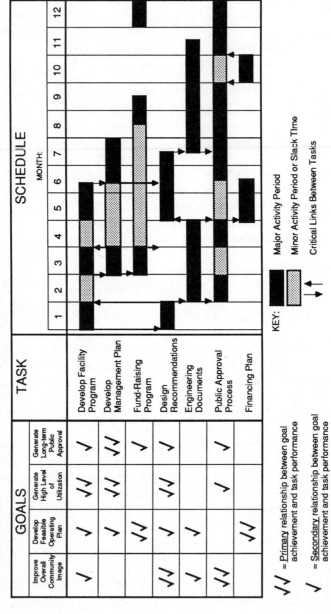

PROPOSED FACILITY DEVELOPMENT PROJECT

FIGURE 6.15 SCHEDULES AND MULTIPLE GOALS (*Explanation*: When a schedule is related to multiple goals it may be too cumbersome to include all of the information on the time-line diagram. A separate but adjacent table, as shown here, is appropriate.)

PROPOSED FACILITY DEVELOPMENT PROJECT

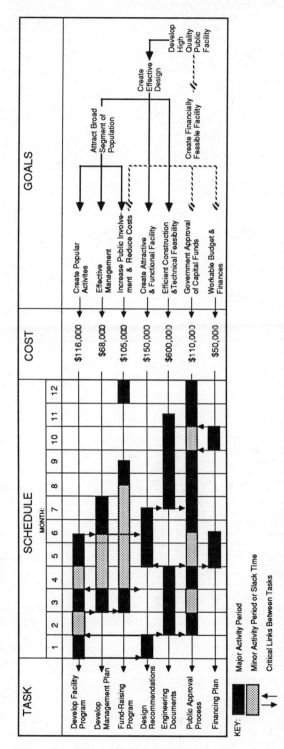

FIGURE 6.16 SCHEDULES AND HIERARCHICAL GOALS (*Explanation:* Goals structured as a "tree" or hierarchy can also be combined with time-line diagrams that show activities and costs. This diagram elaborates the information shown in Figure 6.13)

WORK SCHEDULE FOR COMMUNITY FACILITY PROJECT

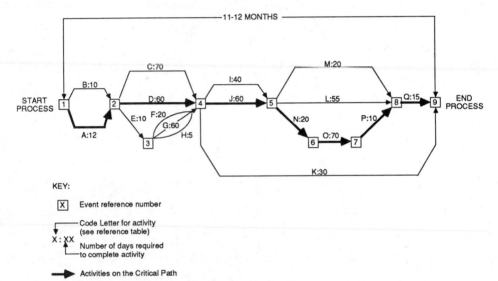

KEY:

| X | Event reference number |

Code Letter for activity
(see reference table)
X : XX
Number of days required
to complete activity

➤ Activities on the Critical Path

CODE LETTER	WORK DAYS	DESCRIPTION	CODE LETTER	WORK DAYS	DESCRIPTION
A	12	REVIEW CURRENT DESIGN CONCEPTS	J	60	MODIFY DESIGN CONCEPTS AS NEEDED
B	10	PREPARE PRELIMINARY PROGRAM	K	30	INITIATE FUND RAISING CAMPAIGN
C	70	CONDUCT PUBLIC MEETINGS AND REVIEW	L	55	CONDUCT PUBLIC MEETINGS AND REVIEW
D	60	DESIGN DEVELOPMENT ACTIVITES	M	20	FINALIZE FINANCING PLAN
E	10	ANALYZE PROGRAM FEASIBILITY	N	20	PREPARE CAPITAL COST ESTIMATE
F	20	DEVELOP MANAGEMENT PLAN	O	70	PREPARE CONSTRUCTION BID DOCUMENTS
G	50	DEVELOP FUND RAISING PLAN	P	10	PREPARE CAPITAL BUDGET
H	5	PREPARE PROGRAM RECOMMENDATIONS	Q	15	COMMITTEE HEARINGS
I	40	PREPARE INITIAL FINANCING PLAN			

FIGURE 6.17 SIMPLE NETWORK DIAGRAMS (**Explanation**: Simple network diagrams communicate general patterns and relationships among activities rather than specific details. This example is based on a situation similar to that shown in Figures 6.12 to 6.16. It portrays a general work program, the number of days to complete each task, the logical links between tasks, and the *critical path* (i.e., the tasks that if not completed as scheduled, will delay the entire project).)

Logical Relations, Geometry, Slack, and Critical Paths

The principle advantage of a network diagram over a bar diagram is that it can accommodate complicated relationships and still be presented effectively. Networks can easily account for single events leading to multiple activities, as well as multiple activities culminating in one event. Like bar diagrams, they can portray activities at very minor levels of detail. Almost any task portrayed on the diagram could be shown as several subtasks, so the analyst must select the most relevant level of generality for the intended audience (Figure 6.18).

The construction of a network diagram often begins with an informal listing of significant activities. The analyst then rates each activity and all links to preceding and subsequent activities, a considerable task in itself. The next step entails the construction of a rough diagram portraying those relationships. This often is visually confusing, with many crossed lines and no clear geometry. Consequently, one of the first presentation issues is the generation of an appropriate visual geometry for the diagram, bearing in mind that the same set of logical relations between activities and events can usually be portrayed in a number of different ways (Figure 6.19).

Theoretically, if two different geometries have the same logical network structure, there is no technical difference between them. In practice, however, different geometries can communicate very different ideas to the audience. In some cases, the same network can be redrawn to portray different phases of activities, the work of different teams, or subsets of objectives. The drawing can emphasize one sequence of activities as central to the project, while other activities appear peripheral. The visual order of the diagram communicates additional information, which should be used to make the presentation more effective.

Although network diagrams can help visualize complex processes, they have one major shortcoming. The length of the lines, unlike those in a bar diagram, is *not* proportional to the length of time required to perform an activity, which is shown only by a number. However, sometimes the geometry of the diagram can be manipulated to show longer time spans as generally longer lines. This is especially appropriate if there are just a few activities that have dramatically different time requirements. For example, if most activities are in the range of 1 to 5 weeks but a few will require 10 weeks or longer, it may be more effective to modify the diagram to emphasize the difference (Figure 6.20).

A useful feature of formal CPM and PERT diagrams is the portrayal of logical sequential relationships that are not linked by ongoing activities. For instance, if activity X cannot begin before information from activity Y is available, and the two activities are conducted by different groups in different parts of a project, the simple transferral of the information may not in itself be an activity worth noting. This type of logical relationship often is shown as a dotted line, drawn from the event at the end of activity Y to the event beginning activity X (Figure 6.21).

Usually, network diagrams emphasize the critical path (as computed using the CPM methodology). This can be technically defined as that sequence of activities in which there is no slack time. Any delay in *any* activity on the critical path automatically leads to a delay in the final completion of the project. For example, activity *A* of an economic development project might represent "assessing property values" while activity *B* refers to "preparing conceptual plans" for property development. Activity *A* might be critical because any delay in assessing property values will delay the final project completion. On the other hand, activity *B* might require three weeks, but if it is not completed for five weeks, it will not delay the remainder of the project schedule.

DIAGRAM A
WORK SCHEDULE

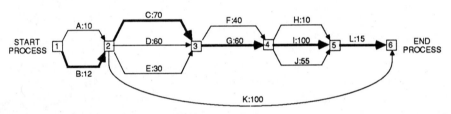

KEY:

 Event reference number

$X : XX$
┌─ Code Letter for activity (see reference table)
└─ Number of days required to complete activity

➤ Activities on the Critical Path

CODE LETTER	WORK DAYS	DESCRIPTION
A	10	REVIEW CURRENT DESIGN CONCEPTS
B	12	PREPARE PRELIMINARY PROGRAM
C	70	CONDUCT PUBLIC MEETINGS AND REVIEW
D	60	DESIGN DEVELOPMENT ACTIVITES
E	30	DEVELOP PROGRAM & MANAGEMENT PLAN
F	40	PREPARE INITIAL FINANCING PLAN
G	60	MODIFY DESIGN CONCEPTS AS NEEDED
H	10	FINALIZE FINANCING PLAN
I	100	PREPARE COST & CONSTRUCTION DOCUMENTS
J	55	CONDUCT PUBLIC MEETINGS & REVIEWS
K	100	PLAN AND INTIATE FUND RAISING CAMPAIGN
L	15	COMMITTEE HEARINGS

FIGURE 6.18 NETWORK SPECIFICITY (**Explanation:** The level of detail that should be shown in a network diagram is a subjective decision by the analyst based on the nature of the problem situation and the audience. For example, the situation portrayed in Figure 6.17 can be made even more general (Diagram A) for a large, less informed audience that will review the data briefly, or the situation can be shown in more detail (Diagram B) for a smaller, more informed audience that will focus on this issue.)

The critical path is often shown as a heavier line on a network diagram to emphasize its importance (Figure 6.22). In some situations, the critical path is used as a basis for making shifts in project resources. If the critical path implies that a project cannot be completed for six months, but there is an externally imposed deadline of five months, the network diagram helps identify where slack resources may be found that could be reallocated to shorten the length of the critical path. In the economic development example, it might be possible to take resources from conceptual planning (activity B) and reallocate them to assessing property values (activity A). Now the situation changes; the planning activity will take more time, thereby using up some of

DIAGRAM B
WORK SCHEDULE

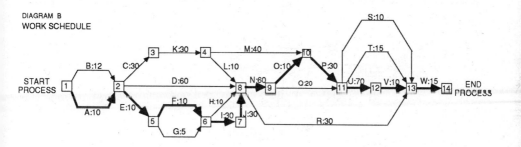

KEY:

[X] Event reference number

X : XX Code Letter for activity (see reference table)
Number of days required to complete activity

➤ Activities on the Critical Path

CODE LETTER	WORK DAYS	DESCRIPTION	CODE LETTER	WORK DAYS	DESCRIPTION
A	10	PREPARE PRELIMINARY PROGRAM	M	40	PREPARE FINANCING PLAN
B	12	REVIEW CURRENT DESIGN CONCEPTS	N	60	MODIFY DESIGN CONCEPTS AS NEEDED
C	30	PUBLIC PRESENTATION OF PROGRAM CONCEPTS	O	10	PREPARE PUBLIC PRESENTATION OF REVISED DESIGN
D	60	DESIGN DEVELOPMENT ACTIVITES	P	30	INCREASE PUBLIC AWARENESS OF RECOMMENDED PROPOSAL
E	10	ANALYZE PROGRAM CONCEPT FEASIBILITY	Q	20	PREPARE DETAILED CAPITAL COST ESTIMATE
F	10	DEVELOP MANAGEMENT OPTIONS	R	30	INTIATE FUND RAISING CAMPAIGN
G	5	PREPARE PROGRAM RECOMMENDATIONS	S	10	FINALIZE FINANCING PLAN
H	10	DEVELOP MANAGEMENT PLAN	T	15	PREPARE STRATEGY FOR PUBLIC HEARINGS
I	30	DEVELOP FUND RAISING OPTIONS	U	70	PREPARE CONSTRUCTION BID DOCUMENTS
J	30	CONDUCT CONFIDENTIAL FUND RAISING SURVEY	V	10	PREPARE BUDGETS AND SCHEDULE FOR HEARING
K	30	PUBLIC PRESENTATION OF PROGRAM RECOMMENDATIONS	W	15	COMMITTEE HEARINGS AND VOTING PROCESS
L	10	ANALYZE PUBLIC RESPONSE			

the slack, while the assessment activities can be completed sooner, thus shortening the overall length of the whole project.

Of course, once the time for an activity is changed, the critical path also may have changed, and must be recomputed. A new critical path may emerge, and new suggestions may be made for reallocating resources (Figure 6.23). It is not unusual to find a presentation in which more than one CPM diagram is needed, each displaying a different trade-off of time and resources.

The inclusion of time/cost/personnel trade-offs in network diagrams is complicated. In simple networks, all three factors can be labelled simultaneously on the diagram, which can be visually confusing, however, when the lines and labels are drawn at varying angles. The effective display of this additional information may again require changing the network geometry.

DIAGRAM A: HYPOTHETICAL FIRST SKETCH

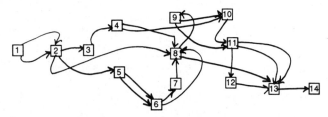

DIAGRAM B: CHANGING GEOMETRIES

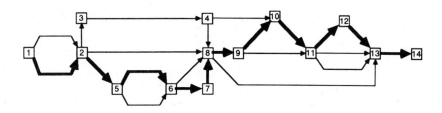

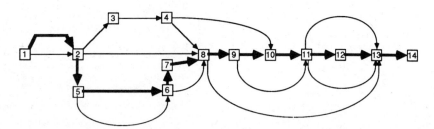

FIGURE 6.19   CHANGING GEOMETRY   (**Explanation:** When a network is first "sketched out" it often looks confusing and chaotic. Diagram A portrays a hypothetical first sketch for Figure 6.18. Diagram B portrays several other visual geometries that match the logical relationships shown in the sketch. In the hypothetical illustration, only the event numbers are shown.)

There should be at least one stretch of horizontal line for each activity, so each line can then accommodate the necessary verbal information (Figure 6.24).

Different personnel allocations may be shown by using several lines in place of one line for each activity. For example, if three teams are involved in a project, a different type of line could be used for each team's involvement in a particular activity. This may violate the principles underlying CPM/PERT methodologies, where technically only one uninterrupted line or flow may connect the same two nodes. Several lines, each representing a different personnel category, breaks this general rule. In practice, it may be more

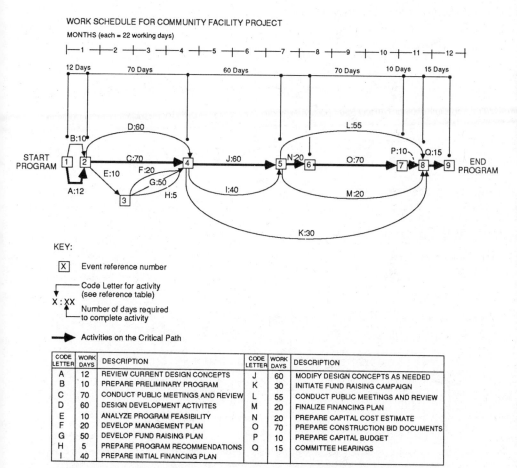

FIGURE 6.20 NETWORKS WITH A VISUAL TIME SCALE (**Explanation**: Most network diagrams emphasize logical relationships between activities but do not visually portray the relative length of activities that are shown on typical time bar or line diagrams. This example shows how a network diagram (identical to the logical relationships shown in Figure 6.17) can be redrawn along a visual time scale, thereby giving the audience some sense of when events will occur.)

effective to show multiple team or personnel activities simultaneously in parallel sequence (Figure 6.25).

If the activities of personnel or organizational units are dramatically different, it may be useful to draw a separate diagram for each. This option could allow for some connecting lines or flows to show the principal interrelationships among otherwise independent work programs (Figure 6.26).

Relevant data on personnel and costs also can be presented in a separate

WORK SCHEDULE FOR COMMUNITY FACILITY PROJECT

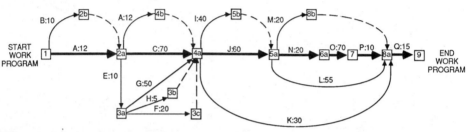

KEY:

| X | Event reference number |

Code Letter for activity
(see reference table)
X : XX
Number of days required
to complete activity

➤ Activities on the Critical Path
→ Other activities
– – ➤ No activity, no time required
(portrays only logical precedence of
tasks and events)

CODE LETTER	WORK DAYS	DESCRIPTION	CODE LETTER	WORK DAYS	DESCRIPTION
A	12	REVIEW CURRENT DESIGN CONCEPTS	J	60	MODIFY DESIGN CONCEPTS AS NEEDED
B	10	PREPARE PRELIMINARY PROGRAM	K	30	INITIATE FUND RAISING CAMPAIGN
C	70	CONDUCT PUBLIC MEETINGS AND REVIEW	L	55	CONDUCT PUBLIC MEETINGS AND REVIEW
D	60	DESIGN DEVELOPMENT ACTIVITES	M	20	FINALIZE FINANCING PLAN
E	10	ANALYZE PROGRAM FEASIBILITY	N	20	PREPARE CAPITAL COST ESTIMATE
F	20	DEVELOP MANAGEMENT PLAN	O	70	PREPARE CONSTRUCTION BID DOCUMENTS
G	50	DEVELOP FUND RAISING PLAN	P	10	PREPARE CAPITAL BUDGET
H	5	PREPARE PROGRAM RECOMMENDATIONS	Q	15	COMMITTEE HEARINGS
I	40	PREPARE INITIAL FINANCING PLAN			

FIGURE 6.21   SHOWING PRECEDENCE RELATIONSHIPS   (***Explanation:*** To facilitate critical path computations, the network diagrams are usually constructed such that there is no more than one single branch or path connecting the same two nodes. This computational requirement leads to the creation of abstract or hypothetical "dummy" variables or activities that are portrayed just to insure the mathematical separation of paths and nodes. This diagram portrays the identical networks as shown in Figure 6.17 and 6.20 with the addition of dummy variables. It also shows how the diagram can be organized to emphasize the critical path.)

table. One dimension of the table can represent the activities in the network diagram, whereas the other represents data on personnel and/or costs. The table can include subtotals as needed. One shortcoming of this approach is that, unlike bar diagrams, which can be visually joined to tables, network diagrams have an unforgiving geometry. This makes it difficult to join networks to other types of diagrams.

WORK SCHEDULE WITH CRITICAL PATH AND SLACK TIME

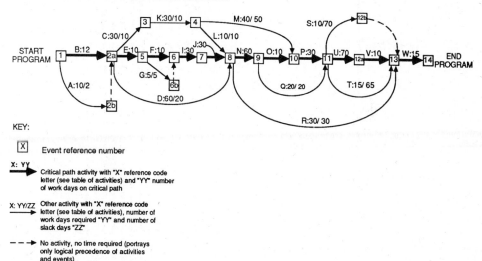

KEY:

$\boxed{X}$	Event reference number
$\longrightarrow$	Critical path activity with "X" reference code letter (see table of activities) and "YY" number of work days on critical path **X: YY**
$\rightarrow$	Other activity with "X" reference code letter (see table of activities), number of work days required "YY" and number of slack days "ZZ" **X: YY/ZZ**
- - →	No activity, no time required (portrays only logical precedence of activities and events)

CODE LETTER	WORK DAYS	DESCRIPTION	CODE LETTER	WORK DAYS	DESCRIPTION
A	10	PREPARE PRELIMINARY PROGRAM	M	40	PREPARE FINANCING PLAN
B	12	REVIEW CURRENT DESIGN CONCEPTS	N	60	MODIFY DESIGN CONCEPTS AS NEEDED
C	30	PUBLIC PRESENTATION OF PROGRAM CONCEPTS	O	10	PREPARE PUBLIC PRESENTATION OF REVISED DESIGN
D	60	DESIGN DEVELOPMENT ACTIVITES	P	30	INCREASE PUBLIC AWARENESS OF RECOMMENDED PROPOSAL
E	10	ANALYZE PROGRAM CONCEPT FEASIBILITY	Q	20	PREPARE DETAILED CAPITAL COST ESTIMATE
F	10	DEVELOP MANAGEMENT OPTIONS	R	30	INTIATE FUND RAISING CAMPAIGN
G	5	PREPARE PROGRAM RECOMMENDATIONS	S	10	FINALIZE FINANCING PLAN
H	10	DEVELOP MANAGEMENT PLAN	T	15	PREPARE STRATEGY FOR PUBLIC HEARINGS
I	30	DEVELOP FUND RAISING OPTIONS	U	70	PREPARE CONSTRUCTION BID DOCUMENTS
J	30	CONDUCT CONFIDENTIAL FUND RAISING SURVEY	V	10	PREPARE BUDGETS AND SCHEDULE FOR HEARING
K	30	PUBLIC PRESENTATION OF PROGRAM RECOMMENDATIONS	W	15	COMMITTEE HEARINGS AND VOTING PROCESS
L	10	ANALYZE PUBLIC RESPONSE			

FIGURE 6.22 THE CRITICAL PATH AND SLACK TIME (**Explanation:** The result of a formal critical path analysis is a computation of the earliest completion time along the critical path plus a computation of the potential slack time for activities not on the critical path. The critical path is the sequence of activities that if delayed imply that the entire process will be delayed—it does not mean activities that are harder to complete or are more expensive. The activities are critical only in terms of minimizing project schedules. The critical path is usually given visual prominence as is shown in the diagram based on Figure 6.18. Slack time, as shown, is defined as the number of days that one noncritical activity can be delayed before it becomes critical, that is, before it will delay the entire project.)

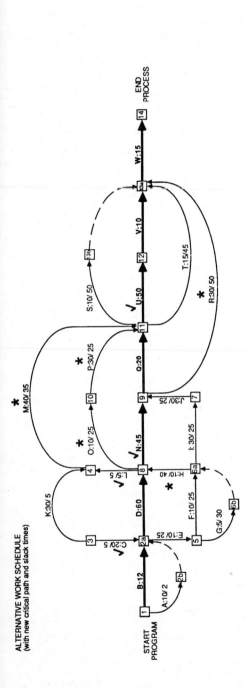

ALTERNATIVE WORK SCHEDULE
(with new critical path and slack times)

CODE LETTER	WORK DAYS	DESCRIPTION
A	10	PREPARE PRELIMINARY PROGRAM
B	12	REVIEW CURRENT DESIGN CONCEPTS
C	20 ✓	PUBLIC PRESENTATION OF PROGRAM CONCEPTS
D	60	DESIGN DEVELOPMENT ACTIVITIES
E	10	ANALYZE PROGRAM CONCEPT FEASIBILITY
F	10	DEVELOP MANAGEMENT OPTIONS
G	5	PREPARE PROGRAM RECOMMENDATIONS
H	10	DEVELOP MANAGEMENT PLAN
I	30	DEVELOP FUND RAISING OPTIONS
J	30	CONDUCT CONFIDENTIAL FUND RAISING SURVEY
K	30 ✓	PUBLIC PRESENTATION OF PROGRAM RECOMMENDATIONS
L	5 ✓	ANALYZE PUBLIC RESPONSE

CODE LETTER	WORK DAYS	DESCRIPTION
M	40	PREPARE FINANCING PLAN
N	45 ✓	MODIFY DESIGN CONCEPTS AS NEEDED
O	10	PREPARE PUBLIC PRESENTATION OF REVISED DESIGN
P	30	INCREASE PUBLIC AWARENESS OF RECOMMENDED PROPOSAL
Q	20	PREPARE DETAILED CAPITAL COST ESTIMATE
R	30	INITIATE FUND RAISING CAMPAIGN
S	10	FINALIZE FINANCING PLAN
T	15 ✓	PREPARE STRATEGY FOR PUBLIC HEARINGS
U	50 ✓	PREPARE CONSTRUCTION BID DOCUMENTS
V	15	PREPARE BUDGETS AND SCHEDULE FOR HEARING
W	15	COMMITTEE HEARINGS AND VOTING PROCESS

KEY:

⊠ Event reference number

X: YY Critical path activity with "X" reference code letter (see table of activities) and "YY" number of work days on critical path

X: YY/ZZ Other activity with "X" reference code letter (see table of activities), number of work days required "YY" and number of slack days "ZZ"

- - -> No activity, no time required (portrays only logical precedence of activities and events)

＊ Activities with new precedence relationships

✓ Shortened time allocations due to changed precedence

FIGURE 6.23 PRESENTING AN ALTERNATIVE SCHEDULE (*Explanation:* Frequently an analysis of the critical path is used to identify activities that are then modified to shorten a project schedule. For example, presume the analysis shown in Figure 3.22 was used as a basis for proposing an alternate schedule in which critical activities would be shortened by reallocating staff time or hiring additional staff and, while the activities remained the same, changes would be made in the logical procedure between tasks. This would lead to a new critical path and might be presented as shown in the diagram.)

206

SCHEDULE AND COSTS FOR FACILITY DEVELOPMENT WORK PROGRAM

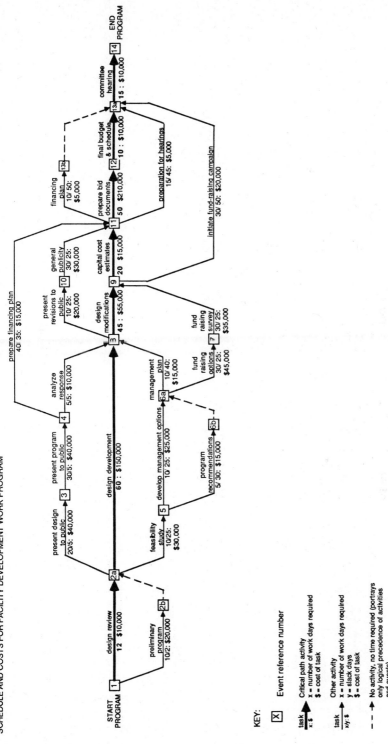

KEY:

☒ Event reference number

task
x: $
Critical path activity
x = number of work days required
$ = cost of task

task
x/y: $
Other activity
x = number of work days required
y = slack days
$ = cost of task

– – ▶
No activity, no time required (portrays only logical precedence of activities and events)

FIGURE 6.24   ADDING INFORMATION TO A NETWORK   (*Explanation:* When more data need to be displayed, the geometry of the diagram should be changed to make the necessary information more legible. One way to do this is to create horizontal lines between all nodes. The diagram uses this technique to add cost data to the information already shown in Diagram 6.23.)

IN-HOUSE AND CONSULTANT ACTIVITIES FOR FACILITY DEVELOPMENT PROGRAM

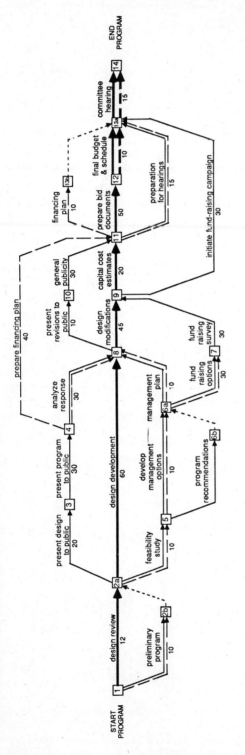

KEY:

☒ Event reference number

▲ Consultant activities on critical
path (with number of work days)

↑ Other consultant activities
(with number of work days)

▲ In-house activities on critical
path (with number of work days)

↑ Other in-house activities
(with number of work days)

⋯ No activity, no time required
(portrays only logical precedence
of activities and events)

FIGURE 6.25 SHOWING SPLIT TASKS ON ONE NETWORK (*Explanation:* In simple cases where there are two primary personnel groups or organizations involved in a project, the network can display the split in activities and associated data. This example is based on the same situation as is shown in Figure 6.24.)

Relation to Objectives

The formal application of CPM and PERT methods always leads to a single end point, implying the achievement of objectives and outcomes inherent in the project. That is, all project objectives and outcomes are visually represented as a single event marking project completion.

In practice, there may be several objectives and outcomes which are achieved along the way, so it may be more effective to separate major objectives and develop a separate network diagram for each (Figure 6.27). The alternative is to use one diagram that denotes several points (possibly multiple end points) signifying achievement of principal objectives. This does not fit the formal requirements of CPM/PERT algorithms, but it does improve the presentation of relevant data. For example, different line symbols (solid, dashed, dotted) can be used to show the sequence of events leading to different objectives (Figure 6.28). Similarly, different line weights can be used to show activities that lead to many objectives (heavy lines), several objectives (medium weight lines), or just one objective (a light line)

Accompanying diagrams, such as tables, can be used to present the relationship of tasks to objectives. This can be useful, although as noted previously, it is difficult to connect the visual geometry of a network diagram to other diagrams or tables.

INTEGRATING WORK PROGRAMS AND DECISIONS: PROCESS DIAGRAMS

Since the advent of computer programming, a new type of diagram has emerged that portrays logical connections, decisions, activities, actions, and outcomes. These diagrams have been used principally to show the structure and logic of computer programs, subprograms, loops, and related issues. However, they can also be used to present data and information concerning project management problems that are interrelated with decision-making problems and outcomes.

These diagrams represent the flow of information through a series of decisions and actions that reach any one of several end points. They look like network diagrams with the one basic exception that the nodes do not represent events, but decisions, activities, outcomes, and other issues. The lines represent the flow of information, where their real purpose is simply to track the sequence of decisions, activities, and outcomes that will occur under different circumstances.

Different symbols are used for different types of nodes. For example, a six sided polygon or lozenge shape usually denotes a decision with two options. One line flows into the node and two flow out. The latter two lines represent the two options, only one of which is taken. In computer programming, decisions usually are binary; that is, there are two choices. In other applications, such as project management, there is no reason to adhere to this

SCHEDULE AND COSTS FOR WORK PROGRAM GROUPS

KEY:

☒ Event reference number

X:YY Critical path activity with "X" reference code letter (see table of activities) and "YY" number of work days on critical path

X:YY Other activity with "X" reference code letter (see table of activities), number of work days required "YY".

→ No activity, no time required (portrays only logical precedence of activities and events)

$$ Cost of activity (in $1,000)

✓ Joint activities between work groups

CODE LETTER	WORK DAYS	DESCRIPTION
A	10	PREPARE PRELIMINARY PROGRAM
B✓	12	REVIEW CURRENT DESIGN CONCEPTS
C	30	PUBLIC PRESENTATION OF PROGRAM CONCEPTS
D	60	DESIGN DEVELOPMENT ACTIVITIES
E✓	10	ANALYZE PROGRAM CONCEPT FEASIBILITY
F	10	DEVELOP MANAGEMENT OPTIONS
G	5	PREPARE PROGRAM RECOMMENDATIONS
H✓	10	DEVELOP MANAGEMENT PLAN
I	30	DEVELOP FUND RAISING OPTIONS
J	30	CONDUCT CONFIDENTIAL FUND RAISING SURVEY
K	30	PUBLIC PRESENTATION OF PROGRAM RECOMMENDATIONS
L✓	5	ANALYZE PUBLIC RESPONSE

CODE LETTER	WORK DAYS	DESCRIPTION
M✓	40	PREPARE FINANCING PLAN
N✓	60	MODIFY DESIGN CONCEPTS AS NEEDED
O	10	PREPARE PUBLIC PRESENTATION OF REVISED DESIGN
P	30	INCREASE PUBLIC AWARENESS OF RECOMMENDED PROPOSAL
Q	20	PREPARE DETAILED CAPITAL COST ESTIMATE
R	30	INITIATE FUND RAISING CAMPAIGN
S	10	FINALIZE FINANCING PLAN
T✓	15	PREPARE STRATEGY FOR PUBLIC HEARINGS
U	70	PREPARE CONSTRUCTION BID DOCUMENTS
V✓	10	PREPARE BUDGETS AND SCHEDULE FOR HEARING
W✓	15	COMMITTEE HEARINGS AND VOTING PROCESS

FIGURE 6.26 SHOWING SPLIT TASKS ON MULTIPLE NETWORKS *(Explanation:* In some situations where several distinct groups or organizations are involved in several tasks it may be advantageous to display the time schedule and associated data on several networks. These three diagrams are all based on the same situation portrayed in Figures 6.23, 6.24, and 6.25.)

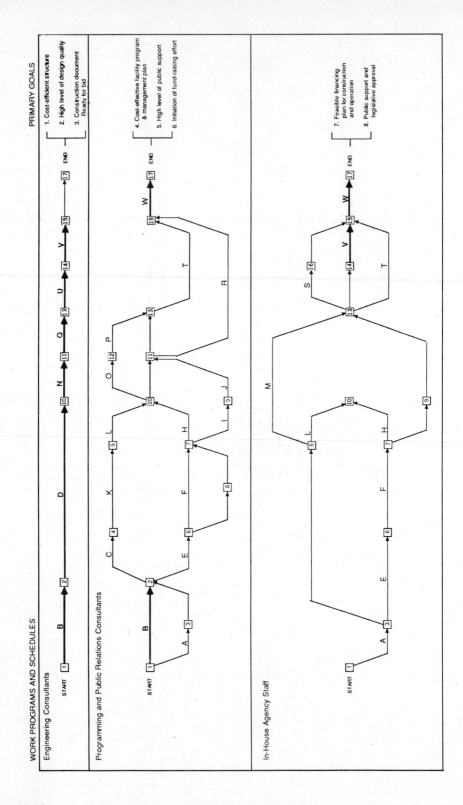

CODE LETTER	WORK DAYS	DESCRIPTION	CODE LETTER	WORK DAYS	DESCRIPTION
A	10	PREPARE PRELIMINARY PROGRAM	M	40	PREPARE FINANCING PLAN
B	12	REVIEW CURRENT DESIGN CONCEPTS	N	60	MODIFY DESIGN CONCEPTS AS NEEDED
C	30	PUBLIC PRESENTATION OF PROGRAM CONCEPTS	O	10	PREPARE PUBLIC PRESENTATION OF REVISED DESIGN
D	60	DESIGN DEVELOPMENT ACTIVITES	P	30	INCREASE PUBLIC AWARENESS OF RECOMMENDED PROPOSAL
E	10	ANALYZE PROGRAM CONCEPT FEASIBILITY	Q	20	PREPARE DETAILED CAPITAL COST ESTIMATE
F	10	DEVELOP MANAGEMENT OPTIONS	R	30	INITATE FUND RAISING CAMPAIGN
G	5	PREPARE PROGRAM RECOMMENDATIONS	S	10	FINALIZE FINANCING PLAN
H	10	DEVELOP MANAGEMENT PLAN	T	15	PREPARE STRATEGY FOR PUBLIC HEARINGS
I	30	DEVELOP FUND RAISING OPTIONS	U	70	PREPARE CONSTRUCTION BID DOCUMENTS
J	30	CONDUCT CONFIDENTIAL FUND RAISING SURVEY	V	10	PREPARE BUDGETS AND SCHEDULE FOR HEARING
K	30	PUBLIC PRESENTATION OF PROGRAM RECOMMENDATIONS	W	15	COMMITTEE HEARINGS AND VOTING PROCESS
L	5	ANALYZE PUBLIC RESPONSE			

FIGURE 6.27 MULTIPLE NETWORKS AND GOALS (*Explanation:* Theroetically, networks should have a single end point representing the achievement of a primary goal. When networks are split (see Figure 6.26), it is often preferable to present distinct goals for each set of tasks or networks as is shown in the diagram.)

IN-HOUSE AND CONSULTANT ACTIVITIES FOR FACILITY DEVELOPMENT PROGRAM

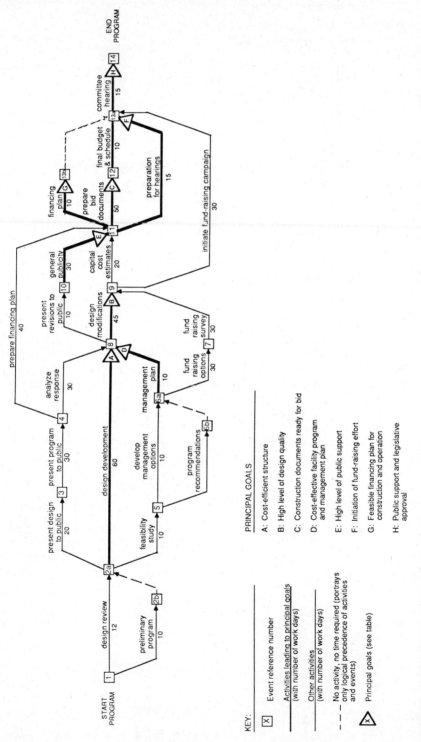

KEY:

☒   Event reference number

—   Activities leading to principal goals
(with number of work days)

—   Other activities
(with number of work days)

---   No activity, no time required (portrays
only logical precedence of activities
and events)

△   Principal goals (see table)

PRINCIPAL GOALS

A: Cost-efficient structure

B: High level of design quality

C: Construction documents ready for bid

D: Cost-effective facility program
and management plan

E: High level of public support

F: Initiation of fund-raising effort

G: Feasible financing plan for
construction and operation

H: Public support and legislative
approval

FIGURE 6.28   MULTIPLE GOALS ON A SINGLE NETWORK   (*Explanation:* In some cases (as is shown in Figure 6.24) it may be preferable to present multiple goals at various points along the network. The analyst must emphasize to the audience that these goals may be equally important and that reaching the "end" of the network is not necessarily the highest priority. The goals shown here are similar to those listed in Figure 6.27.)

constraint. Thus, a decision point could lead to several options similar to the way decision trees are drawn (Figure 6.29).

Most information should be shown at the node points and not along the lines, so where an activity occurs between two decision points, it should be represented in a separate rectangle rather than being drawn along the line (as in a network or bar diagram). Conventionally, such diagrams usually are oriented vertically with the starting point at the top.

Nodes shaped like rectangles usually denote activities. In computer programming, they usually are a set of mathematical operations, whereas in project management terms, rectangles can represent, perhaps, the tasks allocated to different teams.

Other symbols, such as circles and diamonds, represent types of operations and logical structures in computer programming. These, too, can be adapted to a project management situation. Circles can be used to denote outside rather than internal activities, while diamonds can represent cost subtotals, work reviews or personnel assignments. Other symbols can denote the outcomes or various end points which are critical to the presentation. Further graphic indicators can show the *chance points* or events with different probabilities found in decision trees or hierarchies (Figure 6.30).

The principal advantage of using this type of diagram is that it can show more types of information than either networks or decision trees. In fact, it can show information common to both of these techniques. Tasks, for example, can be shown as contingent upon decisions and outcomes that occur along the way.

The diagram can also show *feedback loops* in which certain tasks may be repeated, contingent on subsequent outcomes. Where the results of a task must be reviewed and approved, the approval may not be guaranteed. A feedback loop can show that a positive outcome of the review will lead to the next forward step in the process, while a negative outcome (or disapproval) may lead to a repeat of the prior tasks (Figure 6.31). Regular bar diagrams and networks cannot portray such feedback loops, as they presume all tasks are undertaken once, and that there are no decisions to select one task and to delete another.

Costs can be shown as separate nodes or at each node representing an activity, and cost totals can be shown at various end points contingent on the sequence of tasks and decisions leading to that point (Figure 6.32). The use of different personnel teams can be shown as multiple nodes that all emanate from the same previous task. Outcomes can be shown at several points on the periphery or edge of the diagram.

Thus, this type of process diagram is especially flexible, but does have some shortcomings. The added level of logical complexity, the use of multiple symbols and the abstract quality of the diagram require careful explanation if it is to be effective. Audiences may react negatively if the diagram conveys a rigid, computerlike character to a process which, in fact, may be flexible,

TASKS AND OPTIONS FOR COMMUNITY FACILITY DEVELOPMENT

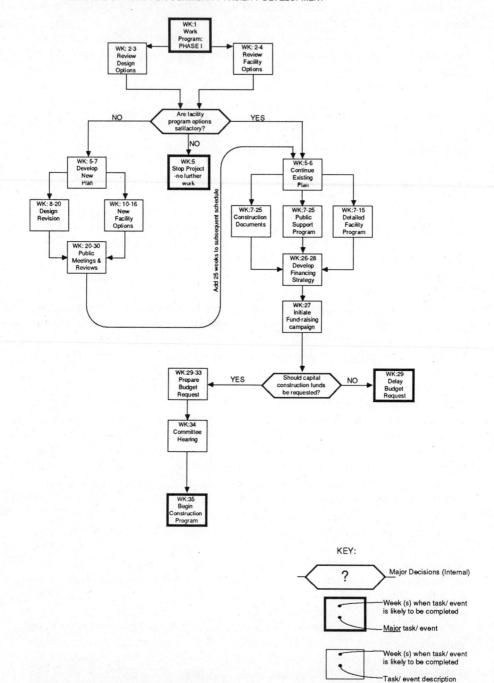

FIGURE 6.29 SIMPLE PROCESS DIAGRAMS (**Explanation:** This diagram is based on situations similar to those portrayed in Figures 6.17 to 6.28. However, this diagram incorporates *decision options*—an item not found on typical CPM diagrams or networks. The statistics presented are simple scheduling data.)

TASKS AND OPTIONS FOR COMMUNITY FACILITY DEVELOPMENT

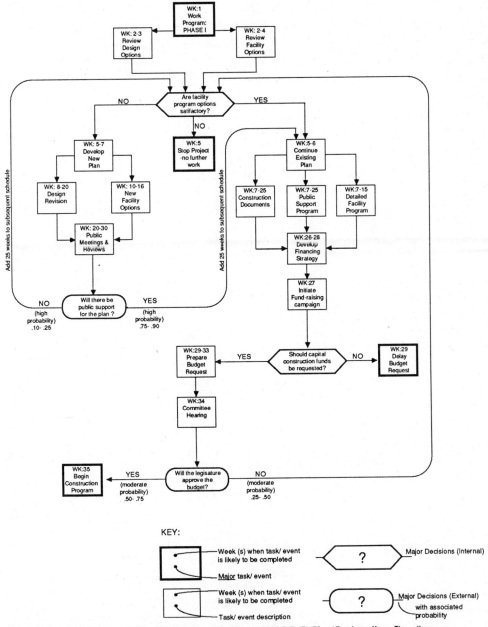

KEY:

Week (s) when task/ event is likely to be completed

Major task/ event

Week (s) when task/ event is likely to be completed

Task/ event description

? Major Decisions (Internal)

? Major Decisions (External)
with associated probability

FIGURE 6.30 COMBINING PROCESS DIAGRAMS WITH CHANCE EVENTS (***Explanation:*** The diagram adds two *chance* nodes, with associated probabilities to the information shown in Figure 6.29.)

TASKS AND OPTIONS FOR COMMUNITY FACILITY DEVELOPMENT

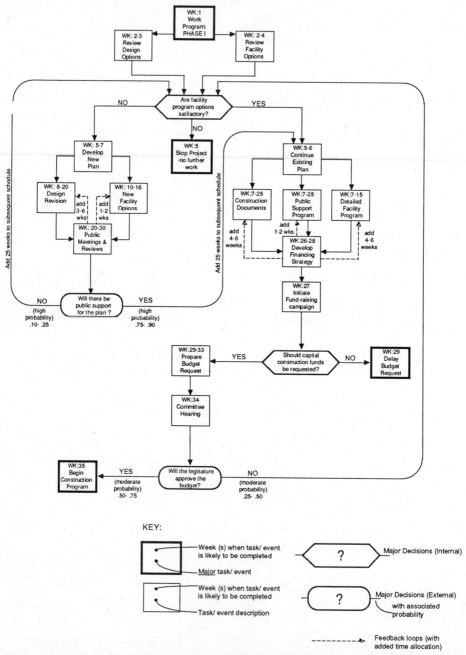

FIGURE 6.31 PRESENTING *FEEDBACK* LOOPS **(*Explanation:*** This diagram indicates several feedback loops, and associated data, for the process shown in Figure 6.30.)

TASKS AND OPTIONS FOR COMMUNITY FACILITY DEVELOPMENT

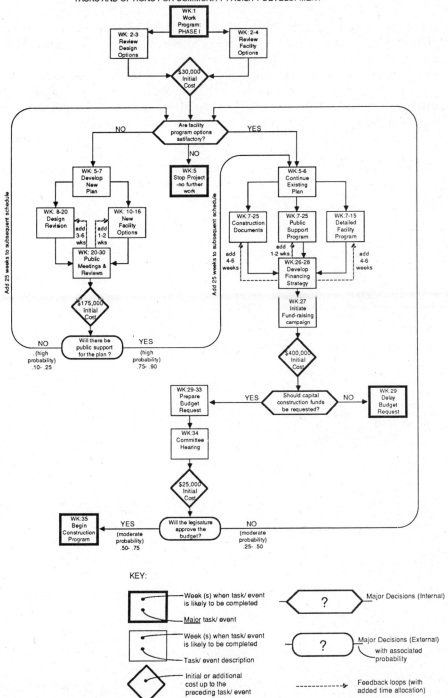

FIGUERE 6.32 EMPHASIZING COSTS IN PROCESS DIAGRAMS (*Explanation:* This diagram includes cost data in separate nodes and thus adds one more factor to the information portrayed in the diagrams shown in Figure 6.29 to 6.31.)

subjective, and humanistic. Furthermore, as in network diagrams, the length of lines and size of nodes are not proportional to actual times and costs, which may be confusing. Another shortcoming lies in the fact that the probability of events can be presented, but not collapsed into expected values and similar figures used in decision trees and matrices. Nevertheless, this type of diagram still has the principal advantage of synthesizing a wide variety of issues and should be used where the analyst believes it will have an appropriate impact.

7

Presentation Style

The previous six chapters have explored the potential techniques available to the analyst in the presentation of audience-oriented statistics. However, in each instance, an inadequate graphic presentation may confuse or undersell the meaning of the work, and lead to audience misunderstanding or disinterest. The purpose of this concluding chapter, therefore, is to forward some basic presentation styles, and to outline some general rules that apply to techniques throughout the book. The following sections, although not comprehensive, do provide a basic foundation of ideas that are intended to make statistical presentations more effective.

ANALYZING THE AUDIENCE

In each case, the presentation design should begin with an analysis of the intended audience in terms of its knowledge, values and expectations. This should be followed by an examination of the specific decisions that will be affected by the presentation, and the issues that must therefore be addressed. Finally, the data that are to be used must be reviewed. Reversing this order is inadvisable, and is not likely to be very effective. Once the structure has been outlined, the details of the presentation can be worked out. Conceptually, the analysis of the data and the analysis of the audience must be interwoven, mutually dependent activities (Figure 7.1). The following sections may be of help in maximizing the delivery of the selected statistical techniques.

CONCLUSIONS FIRST

Good reports and presentations frequently begin with an executive summary or statement of final recommendations. Unfortunately, the practice of quantitative analysis usually is based on the opposite convention, where presentations may begin with questions, assumptions, and methods of inquiry. Even when a presentation begins with an abstract, it may not be sufficiently detailed or specific to communicate adequately the intended information of the report. Therefore, the audience often only finds the answers at the end of a long trail of numbers. In many situations, then, it is more effective to tell the audience where the analysis is headed, and what lies at the end of the maze. There is

A.   EFFECTIVE PRESENTATION PROCESS

STEP 1: ANALYZE **AUDIENCE** CHARACTERISTICS

STEP 2: PINPOINT CRITICAL **ISSUES**

STEP 3: SELECT **PRESENTATION** TECHNIQUE

STEP 4: IDENTIFY RELEVANT **STATISTICS**

STEP 5: COLLECT RELEVANT **DATA**

B.   TYPICAL (BUT INEFFECTIVE) PRESENTATION PROCESS

STEP 1: COLLECT AVAILABLE **DATA**

STEP 2: SELECT **STATISTICS**

STEP 3: SELECT **PRESENTATION** TECHNIQUE

STEP 4: PINPOINT CRITICAL **ISSUES**

STEP 5: ANALYZE **AUDIENCE** CHARACTERISTICS

FIGURE 7.1 PRESENTATION PROCESS (*Explanation:* Table A lists the broad conceptual steps that should be used to develop an effective presentation. Table B illustrates a typical, but ineffective process in which the analyst often presents statistics and issues that do not fit the intended audience. The process must begin with an analysis of audience characteristics, such as the number and type of people, their level of knowledge, their special interests and concerns, and the amount of time for the presentation.)

nothing unprofessional or unreasonable about beginning a presentation with the analytic conclusions if this aids audience comprehension.

The tradition of placing analytic conclusions at the end of a statistical presentation probably derives from a perceived attitude toward rationality, where analysts may believe that an audience of rational people will accept a conclusion only after they understand each piece of evidence and each step in the supporting argument. They may experience some anxiety that an audience may reject a conclusion if it is stated at the outset based on their personal values and beliefs and before they have had a chance to explain the data. This presumes that stating conclusions at the end avoids such anxieties over premature audience judgments.

In practice, however, audience behavior may vary significantly. Audience members may guess at conclusions at the start of a presentation, even when they are not stated. Readers may begin by looking at the end of the report, or may hear the whole argument and still reject the conclusions based on personal values.

If the data and arguments emerge gradually, any uneasiness and uncertainty that may occur when audience members do not know where the presentation is leading should fade. Moreover, they can judge the assumptions and techniques in their own minds relative to the conclusions, and focus their attention more effectively on points they find questionable. If an audience does not know where analysis will lead, they may, in fact, be suspicious of all the computations and data, or simply lose the thread of the argument and become confused or disinterested.

Most audiences appreciate the candor, simplicity, and direction that come from an initial statement of conclusions. Audience members who react negatively to this approach would probably react negatively even if conclusions are postponed to the end.

By beginning at the end, the presentor has a much better chance of holding the audience's attention. Moreover, the presentor has a framework in which to emphasize analytic issues that has a more direct bearing on the results. This can make the presentation quicker, more specific, and more relevant to the audience.

TITLES AND SUBTITLES

In presenting statistics, one of the most important single decisions is the creation of titles for the subject matter. The numbers are meaningless without a logical, analytic framework in which they are perceived. The labelling of statistical illustrations too often derives from the rules governing statistical computations rather than from an appreciation of the intended audience.

The traditional criterion for creating a title is accuracy. This often leads to a long string of words that precisely and comprehensively define the types of numbers contained in a statistical array. However, an accurate title that is not fully understood by the audience is, in fact, misleading or distortive of the information that it is intended to reference. Accuracy depends not just on statistical issues, but equally on the communication between the statistical presenter and the audience.

The criteria for creating a good title include the swiftness of audience comprehension, an absence of overly technical concepts, and a clear relationship to the argument or context in which the statistics are presented. Titles should be like slogans—short, brief phrases that attract immediate attention, are easy to remember, and are easy to relate to a goal, viewpoint, or significant

	HYPOTHETICAL, LONGER MORE EXPLANATORY TITLES TYPICAL OF MANY PRESENTATIONS	SHORT "PUNCHY" TITLES
EXAMPLE A	MEDIAN INCOMES FOR FAMILIES, BY TYPE OF COMMUNITY, FOR 1940 THROUGH 1980	FAMILY INCOMES BY AREA (1940-1980)
EXAMPLE B	PROJECTED INCOMES FOR FAMILIES BY TYPE OF COMMUNITY FOR 1990 THROUGH 2020	INCOME PREDICTIONS: THE NEXT THIRTY YEARS
EXAMPLE C	CRITICAL PATH ANALYSIS, TASKS AND SLACK TIME FOR TWELVE MONTH WORK PROGRAM	NEXT YEAR'S WORK PROGRAM
EXAMPLE D	EFFECTIVENESS/ COST RATIOS FOR ALTERNATIVE DEVELOPMENT PROJECTS FOR FISCAL YEAR '88/ '89	PROJECT RECOMMENDATIONS FOR NEXT YEAR
EXAMPLE E	UNEMPLOYMENT RATES AND FREQUENCIES FOR MAJOR INDUSTRIES AND COMMUNITIES FOR SIX-COUNTY AREA	REGIONAL UNEMPLOYMENT STATISTICS
EXAMPLE F	ALTERNATIVE ACTIONS, PAYOFFS AND RECOMMENDED DECISIONS FOR THE HEALTH FACILITY PROGRAM	HEALTH FACILITY RECOMMENDATIONS
EXAMPLE G	RELATIVE WEIGHTED VALUES FOR COMPARING ALTERNATIVE TRANSPORTATION PROJECTS	POINT SCORES FOR TRANSPORTATION PROJECTS

FIGURE 7.2 SHORTENING TITLES (**Explanation:** This table shows examples of longer, more explanatory titles versus shorter summary titles. Typcially the shorter versions are easier to understand and more effective in presentations. The necessary qualifications for a table or diagram can be presented orally or in footnotes and in supplementary explanations.)

issue. When the audience reads the title of a table, it should be able to infer easily how that table may relate to the larger issues and arguments (Figure 7.2). They should not wonder why the table is there, or be unable to appreciate the data without a lengthy explanation.

The first step in creating an effective title is to empathize with the types of audience members. Their possible interest in the statistics should be questioned, and some key words associated with those interests should be identified. Those key words should then be used to create a rough title. The title should then be shortened, if possible, and any jargon should be eliminated. Titles should be tested with colleagues and reworded as necessary to fit both the interest of the audience and the statistics being presented.

Five to ten minutes spent on developing a title is not a waste of time if, without a good title, the potential value or meaning of the statistics is lost. There usually is a limited amount of time available to design a statistical presentation, but there is nothing unprofessional or irrational about spending more time on titles and less on including more statistics.

DEFINITIONS AND ASSUMPTIONS

In addition to titles, the definitions of terms, categories, and similar items of information are critical. Terms and categories usually are defined to fit the existing structure of the data, but the reverse possibility of reorganizing the data to fit new definitions is rarely considered, and definitions may be more effective if created to fit the audience and the presentation first.

The choice of which way to define and present data is a subjective judgment that should be based on understanding of the validity and reliability of the data, the relevance of the data, and the character of the audience. It would be incorrect to presume that data sets that have greater validity and lower margins of error are automatically more relevant.

Therefore, definitions and assumptions should be modified to alter the character of the presentation. This is a difficult task requiring creativity and discipline, but it should be viewed as an integral part of any statistical presentation, and extra effort should always be devoted to exploring the impact of alternative definitions and assumptions. If a new definition makes data more relevant but potentially less accurate, the analyst should seriously consider the trade-off, and not dismiss it casually as an anti-intellectual or irrational response.

A well-known example of how definitions and assumptions alter data occurs in the way official unemployment rates are calculated. In some cases, the rate is calculated by including in the numerator only those workers receiving government unemployment benefits. Some definitions are expanded such that the calculations also include workers who no longer receive benefits and who are *discouraged* from seeking employment. The statistics can be further modified by defining *worker* to include anyone who is eligible for work rather than just those who have previous records of full-time employment (Figure 7.3).

Operational definitions and assumptions can also be used to combine data categories. For example, a data set may include 10 categories of people. Each category of people will be affected by one of two different programs (*A* and *B*) under discussion. Instead of displaying 10 data categories, the data can be aggregated in group *A* and group *B*, where each group is operationally defined as those persons who will be affected by the program (Figure 7.4).

Numerical data may be transformed into ordinal data by using operational definitions. For instance, the categories of *high*, *moderate*, and *low* can be used to rate the costs and benefits of different actions. Each term can be operationally defined as a precise numeric range, or a combined set of smaller ranges.

Analysts often avoid such techniques because they involve modifying precise and discrete data and interjecting subjective judgments and values. This frequently diminishes the apparent validity and reliability of the data. It also requires explanatory arguments that cannot be based solely on computational issues and the mechanics of measurement. In other words, it places the analyst in a new role with new responsibilities to the audience, a role for which most analysts have not been educated, but one that is likely to have effective results.

COMPUTATION ILLUSTRATIONS

Statistical techniques involve computations, formulas, and specialized vocabularies with which the audience may be unfamiliar. Even when audience members are knowledgeable about a technique, they may be unfamiliar with its application in the situation being discussed. They may have forgotten parts of the technique they once knew, or may have learned a similar technique, which prompts uncertainties about the actual technique being used. Some persons may be embarrassed about asking questions, or may be intimidated by the sophistication of the techniques.

Analysts often rely subconsciously on such phenomena to avoid detailed, lengthy explanations of computational methods and applications. In fact, such statistical issues may be of minor relevance to the audience. Nevertheless, these situations usually result in the audience accepting statistical results as an act of faith rather than by reason and logic. This, in turn, adds to the image of quantitative analysis as mystifying or technocratic.

In the worst case, the audience may reject the conclusions—even when they are very reasonable—simply because they do not understand the process. It may also become a political decision, where one group endorses the conclusions just because they like them, and another group reacts negatively

TABLE A.   UNEMPLOYMENT RATE *

	CITY	SUBURBS
YOUTH (UNDER 18)	2.0	1.0
ADULT MALES	1.8	2.0
ADULT FEMALES	1.2	1.0
TOTAL	5.0	4.0

* PERSONS PER 1,000 CURRENTLY RECEIVING UNEMPLOYMENT BENEFITS

TABLE B.   UNEMPLOYMENT RATE *

	CITY	SUBURBS
YOUTH (UNDER 18)	2.8	1.3
ADULT MALES	4.0	2.3
ADULT FEMALES	1.8	1.3
TOTAL	8.6	4.9

* PERSONS PER 1,000 CURRENTLY RECEIVING UNEMPLOYMENT BENEFITS OR WHO ARE "DISCOURAGED" UNEMPLOYED WORKERS NO LONGER SEEKING JOBS

TABLE C.   UNEMPLOYMENT RATE *

	CITY	SUBURBS
YOUTH (UNDER 18)	3.0	1.7
ADULT MALES	5.0	3.3
ADULT FEMALES	4.0	2.7
TOTAL	12.0	7.7

* NUMBER OF PERSONS PER 1,000 (BASED ON A TELEPHONE SURVEY ESTIMATE) WHO DESIRE WORK AND ARE CAPABLE OF WORKING BUT WHO ARE NOT EMPLOYED OR WHO ARE UNDEREMPLOYED

FIGURE 7.3   EMPHASIZING DATA DEFINITIONS   (**Explanation:** The Tables A, B, and C illustrate how the same apparent issue can be measured three different ways, each producing a different result. In such cases, the operational definition of the data should be given special emphasis.

TABLE A. HEALTH SCREENING STATISTICS

	NUMBER OF PERSONS SCREENED	COST
CITY SCHOOL CHILDREN	40,000	$200,000
SUBURBAN SCHOOL CHILDREN	25,000	$100,000
ELDERLY CITY RESIDENTS	18,000	$90,000
ELDERLY SUBURBAN RESIDENTS	25,000	$100,000
DOWNTOWN OFFICE WORKERS	60,000	$225,000
PUBLIC EMPLOYEES	42,000	$100,000
OTHER UNION WORKERS	45,000	$240,000
OTHER NON-UNION WORKERS	20,000	$168,000
UNIVERSITY STUDENTS	10,000	$50,000
OTHERS	55,000	$220,000
TOTAL	340,000	$1,493,000

TABLE B. HEALTH SCREENING STATISTICS

	NUMBER OF PERSONS SCREENED	COST
CITY SCREENING PROGRAMS [1]	133,000	$665,000
STATE SCREENING PROGRAMS [2]	207,000	$828,000
TOTAL	340,000	$1,493,000

[1] INCLUDES CITY SCHOOL CHILDREN, ELDERLY CITY RESIDENTS, DOWNTOWN OFFICE WORKERS, PUBLIC EMPLOYEES, AND UNIVERSITY STUDENTS

[2] INCLUDES SUBURBAN SCHOOL CHILDREN, ELDERLY SUBURBAN RESIDENTS, UNION AND NON-UNION WORKERS AND OTHER AREA RESIDENTS NOT INCLUED IN CITY PROGRAM.

FIGURE 7.4 DEFINITIONS THAT SUMMARIZE (*Explanation:* In this illustration the data in Table A are summarized more effectively in Table B by defining new categories of data.)

because they do not. Neither group responds on the basis of understanding, the analyst becomes frustrated, since the work has not been fully appreciated, and the audience views the analysis as a failure.

Given the politics of many decision-making processes, some of these problems may occur regardless of an effective presentation. However, proper explanation of statistical procedures can, at the very least, focus discussion on the significant issues rather than the emotional appeal of statistical results.

In most instances, only a few illustrations of statistical processes or terms are needed. They should be full explanations, using simple numbers and as little jargon as possible. Each statistic should be interpreted, so it may be useful to use the statistic in a simple sentence that illustrates its meaning (Figure 7.5). Sometimes, two kinds of verbal examples might be helpful to fully illustrate its meaning (Figure 7.6). One example would use the statistic in a technically correct manner with specialized vocabulary, whereas the other illustration

TERMS:	TECHNICAL DEFINITIONS
MEDIAN INCOME	The number for which 50% of the recorded income values are **higher** and 50% are **lower**
CRITICAL PATH	The logical sequence of activities which, if delayed will necessarily **delay the earliest possible time** of completion for the project.
ERROR RANGE	The upper and lower numerical limits for a given estimate which have a high probability (usually 90% or more) of not being exceeded.
EFFECTIVENESS/ COST RATIO	The quantitative estimate of the effects of an action divided by its cost (usually expressed in dollars).
SEASONALLY ADJUSTED UNEMPLOYMENT RATE	The actual quarterly unemployment rate increased (or decreased) according to the historical pattern of rates for that quarter.
MINIMIZING MAXIMUM LOSS	A criterion for selecting an action from among a set of actions, in which, first, the **worst possible** outcomes for each action are identified and, second, the action with the **best** of these identified outcomes is selected.
WEIGHTED VALUE	A mathematical formula for combining the numerical **weight** assigned to a factor or variable and the numerical **value** of each item contained in that factor (usually expressed as the product of the weight times the value of the associated variable.

FIGURE 7.5 TECHNICAL DEFINITIONS (**Explanation:** For more knowledgable audiences, the analyst may want to present more technical definitions of statistical concepts. These examples are not intended as complete statistical definitions, which often require mathematical equations, but instead as precise statements that minimize jargon.)

would use common everyday terms that may not be technically accurate, but that convey the basic principle.

These types of explanations can be placed at the end of a presentation. They should be mentioned, however, at an earlier opportunity so that the audience is aware of them and can study them as needed. Sometimes, computations of illustrations should be given special prominence. This may be necessary when there is a special type of statistic that is used repeatedly and that is relatively abstract, such as an index, ratio, or coefficient (Figure 7.7). It should always be remembered that if the audience does not understand the statistics at the outset, the entire presentation can be lost.

TERMS:	NON-TECHNCIAL DEFINITIONS
MEDIAN INCOME	Half the people make more money and half make less.
CRITICAL PATH	The list of tasks that must be finished on time.
ERROR RANGE	A conservative guess as to how high and how low the actual number might be
EFFECTIVENESS/ COST RATIO	Whether or not one project is a "good buy" compared to another one.
SEASONALLY ADJUSTED UNEMPLOYMENT RATE	Changing actual unemployment rates to reflect usual seasonal changes (such as higher rates every fall season).
MINIMIZING MAXIMUM LOSS	Basing decisions on a pessimistic viewpoint by avoiding the worst outcomes.
WEIGHTED VALUE	A scoring system which should give more points to more important issues.

FIGURE 7.6 NONTECHNICAL DEFINITIONS (*Explanation:* Some presentations require an explanation of statistical terms using simple language. The definitions correspond to those in Figure 7.5. These definitions may not be as precise, but they provide a general sense of the terms.)

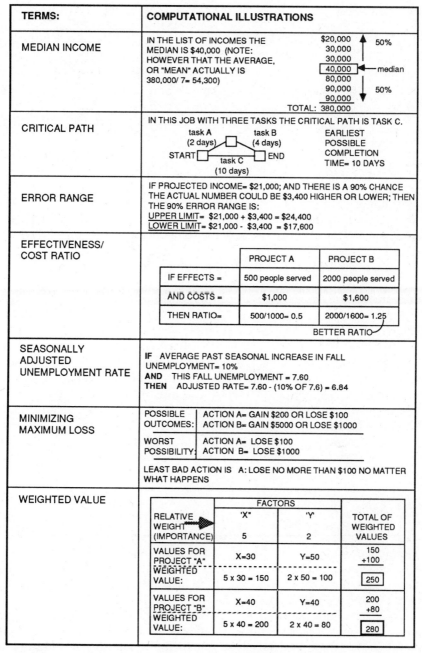

TERMS:	COMPUTATIONAL ILLUSTRATIONS
MEDIAN INCOME	IN THE LIST OF INCOMES THE MEDIAN IS $40,000 (NOTE: HOWEVER THAT THE AVERAGE, OR "MEAN" ACTUALLY IS 380,000/ 7= 54,300) $20,000 — 50% 30,000 30,000 40,000 ← median 80,000 90,000 — 50% 90,000 TOTAL: 380,000
CRITICAL PATH	IN THIS JOB WITH THREE TASKS THE CRITICAL PATH IS TASK C. task A (2 days) task B (4 days) EARLIEST POSSIBLE START — task C (10 days) — END COMPLETION TIME= 10 DAYS
ERROR RANGE	IF PROJECTED INCOME= $21,000; AND THERE IS A 90% CHANCE THE ACTUAL NUMBER COULD BE $3,400 HIGHER OR LOWER; THEN THE 90% ERROR RANGE IS: UPPER LIMIT= $21,000 + $3,400 = $24,400 LOWER LIMIT= $21,000 - $3,400 = $17,600
EFFECTIVENESS/ COST RATIO	

	PROJECT A	PROJECT B
IF EFFECTS =	500 people served	2000 people served
AND COSTS =	$1,000	$1,600
THEN RATIO=	500/1000= 0.5	2000/1600= 1.25

BETTER RATIO

TERMS:	COMPUTATIONAL ILLUSTRATIONS
SEASONALLY ADJUSTED UNEMPLOYMENT RATE	IF AVERAGE PAST SEASONAL INCREASE IN FALL UNEMPLOYMENT= 10% AND THIS FALL UNEMPLOYMENT = 7.60 THEN ADJUSTED RATE= 7.60 - (10% OF 7.6) = 6.84
MINIMIZING MAXIMUM LOSS	POSSIBLE OUTCOMES: ACTION A= GAIN $200 OR LOSE $100 ACTION B= GAIN $5000 OR LOSE $1000 WORST POSSIBILITY: ACTION A= LOSE $100 ACTION B= LOSE $1000 LEAST BAD ACTION IS A: LOSE NO MORE THAN $100 NO MATTER WHAT HAPPENS
WEIGHTED VALUE	

	FACTORS		
	'X'	'Y'	TOTAL OF WEIGHTED VALUES
RELATIVE WEIGHT (IMPORTANCE)	5	2	
VALUES FOR PROJECT "A"	X=30	Y=50	150 +100
WEIGHTED VALUE:	5 x 30 = 150	2 x 50 = 100	250
VALUES FOR PROJECT "B"	X=40	Y=40	200 +80
WEIGHTED VALUE:	5 x 40 = 200	2 x 40 = 80	280

FIGURE 7.7 COMPUTATIONAL ILLUSTRATIONS (**Explanation:** The table shows hypothetical computations that should be used to further explain the terms defined in Figures 7.5 and 7.6.)

GRAPHIC FOCUS, COMPOSITION, AND IMPLEMENTATION

Professional analysts rarely have previous training or experience in graphic design, and yet graphic skill is a major ingredient of a successful presentation. Therefore, consultation with graphic experts is desirable if time and money are available, although a general understanding of basic principles should be developed by analysts to facilitate the process of presentation.

Two of the more significant design issues likely to be confronted are focus and composition. Graphic designers are aware of the need to match the most important information to the most visually dominant graphic elements, which may be titles, numbers, or other diagrammatic relationships. However, the graphic design issues are far more complex if they are to be wholly effective. For example, there may be one major focus, supplemented by a hierarchy of issues of descending significance. The presentation should be matched to those issues so the the importance of the secondary issues is not overly diminished.

The overall composition of the graphic presentation must also be designed effectively. The tendency to cram too much information into one image to ensure maximum delivery may confuse or distort the final effect. Establishing graphic balance, hierarchy and pleasing composition is a major task and requires careful consideration and planning from the outset of the project.

Some graphic presentations will involve the use of a variety of techniques enabling reproduction, reduction, enlargement, and so on. The graphic effect achieved by a particular process may vary from the original and produce a visual image different from that which was initially intended. If the analyst is personally preparing the presentation, care should be taken to ensure that the final graphics project the information in the hierarchy intended, and some prior experimentation or at least inquiry into the processes is advisable before commitment is made to a particular technique.

REPRESENTATIONAL GRAPHICS

In some situations, it may be useful to give the presentation an exotic flavor by animating or emphasizing the graphics in some way. Many examples of this technique can be found in national magazines and on news programs, in which statistical displays are depicted in a way that visually respond to the subject matter under analysis (Figure 7.8).

The technique may appear superficial or tasteless if used on technically oriented or serious presentations, but may be effective in attracting audience attention at a more general level, where the image clearly indicates the intention and subject area behind the statistics without further study.

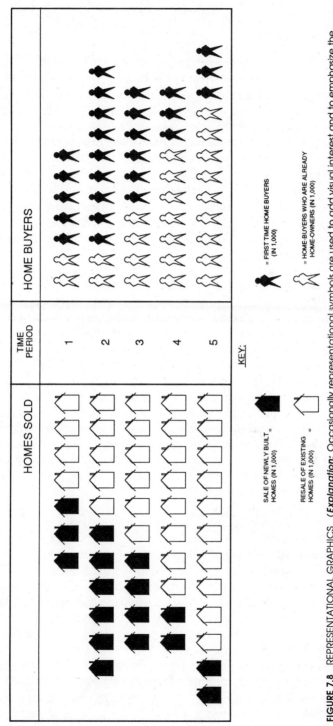

FIGURE 7.8 REPRESENTATIONAL GRAPHICS (*Explanation:* Occasionally representational symbols are used to add visual interest and to emphasize the data.)

233

COMPUTER GRAPHICS

The development of computer graphics over the past few years has revolutionized the presentation process in many areas, and has given the analyst an almost infinite range of options to use in the depiction of statistical information. Of course, new skills are necessary to utilize the available software, and the analyst may have to learn a whole new range of techniques to maximize the potential of the computer. Similarly, the breadth of possibilities laid open by contemporary technology may sometimes make it difficult to chose between the available programs. When computers are introduced, the attainment of clear presentation objectives may be even more difficult to establish. Nevertheless, the flexibility and immediacy demonstrated by many computer graphic programs make them an invaluable tool in the presentation process, and analysts should investigate the potential of various available programs for integration in their work.

The images generated by computer graphics usually can be either printed and copied as part of a larger text (either in report or handout form) or converted to slide or videotape format. In the latter instance, the images generated can be worked together in a sequence to provide audiences with smooth, logical displays of information that can dramatize changes over time or incremental changes brought about by fluctuating variables. This display method is particularly appropriate to the "serial" images discussed in Chapter One.

A number of computer graphic packages are available that enable statistical development and then offer several diagramming options to display the results. In addition to these, there are other types of software packages designed for commercial artists rather than statisticians that, by virtue of their graphic flexibility, may be more effective in some presentation situations. Such programs typically give the user a wide range of type styles, enabling variation and emphasis in large areas of text. Other commercial artwork packages offer the generation of freehand sketching and diagramming, and the addition of graphic elements (arrows, shading, circles, for example) to allow the presentor to embellish and accentuate the data. These programs may allow the user to draw, write, or diagram over a video image of a table, chart, or map and thereby introduce another level of information to the audience.

Computer graphic programs also are available that can be used in mapping exercises and produce relatively sophisticated results. Perhaps the best example of this can be seen on nightly television weather reports, where prevailing and past meteorlogical data are coded into understandable images, often incorporating a time-lapse element.

Some graphic software provide a supplement to general statistical analysis programs that involve the testing of hypotheses, correlations and regression analyses, analyses of variance, and other more sophisticated methods. The analyst should be careful when using these techniques to ensure that the specific computational formulas being used are fully understood, and that the programs being used are entirely consistent. Any differences in the methods of computing of the programs may result in erroneous information being generated.

In conclusion, the range of options available in the presentation of statistics by computer graphics is limited only by the prevailing technology and the interest and/or knowledge of the presentor. Careful investigation of the available relevant programs is a worthwhile task, although care should be taken to choose packages that are within the understanding of the user and that are compatible with the presentation problems and especially the target audiences likely to be encountered.

Appendix

The following list contains the basic references that the reader should consult when seeking more detailed information about the computational procedures and assumptions that are used in preparing statistical analyses and related data for presentation.

GENERAL TEXTS FOR BASIC STATISTICS

1. *Computational Handbook of Statistics*, James L. Brining and B.L. Kintz, Scott, Foresman and Company, 1968, 269 pp.

   This text has a strong pragmatic orientation. It avoids theory to focus on "how-to-do-it." It is oriented primarily toward experimental or inferential statistics.

2. *A Basic Course in Statistics*, 3rd ed., Theordore R. Anderson and Marris Zeldritch, Jr., Holt, Rinehart and Winston, Inc., 1975, 372 pp.

   This text is oriented toward sociology and related social sciences. Its strength is its relatively greater focus on simple descriptive statistics with numerous examples. It contains a few pages on the proper use of statistics.

3. *Elementary Social Statistics*, Kenneth J. Downey, Random House, 1975, 310 pp.

   This text also focuses on simple statistics. Two of its chapters are partially relevant to communication/presentation issues: the first chapter discusses the uses of statistics and the third chapter discusses tables and graphs.

SPECIALIZED TEXTS

1. *The Visual Display of Quantitative Information*, Edward R. Tufte, Graphics Press, 1983.

   This is an elegant analysis of the range of graphic techniques that are used to present data. It includes many historical examples as well as contemporary illustrations. It also shows the effective use of color and other more elaborate, albeit expensive, production techniques. The author presents the work in the context of a conceptual framework that explores the intellectual and theoretical basis for graphic presentation of statistics.

2. *Basic Methods of Policy Analysis and Planning*, Carl V. Patton and David S. Sawicki, Prentice-Hall, 1986, 450 pp.

   This book shows how basic statistics can be used in policy sciences and elaborates many of the related methods and techniques used by planners. It contains numerous examples of how data can be presented in relation to goal formulation, projections, evaluation, and survey research. It also contains several pragmatic case studies.

3. *Urban Planning Analysis: Methods and Models*, Donald A. Krueckeberg and Arthur L. Silvers, John Wiley & Sons, Inc., 1974, 486 pp.

   This text is targeted at urban planners and related professionals. It contains some statistical analysis as well as other forms of quantitative analysis related to scheduling, goal formulation, program evaluation, population projections, and related subjects. The examples and illustrations are pragmatic and relevant to practitioners.

4. *Design for Decision*, Irwin D.J. Bross, The Free Press, 1953, 276 pp.

   This is an excellent introductory text for statistical decision-making techniques. It indirectly provides ample information on how to present statistical arguments in decision-making situations. It is simple and pragmatic.

5. *Operations Research*, Frederick S. Hillier and Gerald J. Lieberman, 1967, Holden-Day, Inc., 1,979 pp.

   This is an excellent text for both introductory and advanced reading. Although it contains numerous examples showing network diagrams used in complex mathematical problems, it also illustrates simple network and flow diagrams based on less complex mathematics, which can be used for general audiences. In particular, the chapter on PERT-CPM techniques is useful.

6. *Decision Analysis: Introductory Lectures on Choices Under Uncertainty*, Howard Raiffa, Addison-Wesley, 1970, 312 pp.

This book contains a detailed, mathematical approach to incorporating subjective values, payoffs, costs, and probabilities in hierarchical decision analysis. It has many examples of decision trees and how they are used to present information. It is, however, oriented toward decision-making situations that conform to given mathematical assumptions that are not always present in complex day-to-day situations.

7. *Nonparametric Statistics for the Behavioral Sciences*, Sidney Siegel, McGraw-Hill, 1956, 312 pp.

This text is oriented toward psychologists and researchers in related disciplines who need a nuts-and-bolts guide to computing relatively specialized statistics.

8. *Methods of Regional Analysis*, Walter Isard, MIT Press, 1960, 784 pp.

This is a classic text that includes a broad range of examples in presenting economic data. It includes projection techniques (such as regression, logistics, and Gompertz curves), tables showing economic conditions, a wide array of indices, coefficients, and ratios designed to measure specific economic features, and diagrams showing economic flows and related concepts.

ADVANCED TEXTS FOR GENERAL STATISTICS

1. *Social Statistics*, Hubert M. Blalock, Jr., McGraw-Hall, 1972, 583 pp.

This is an excellent text for social science students, primarily in sociology, political science, and urban planning. It offers some brief general advice on presentation in the first chapter. It contains many examples, illustrating statistical theory and computation. It describes underlying assumptions and rules for analysis. Recently, a second edition was printed.

2. *Statistical Methods*, George W. Snedecor and William G. Cochran, 6th ed., Iowa State University Press, 1967, 593 pp.

This is a long-standing text, first published in 1937. It is well-recognized and regarded. It is comprehensive in terms of the statistical techniques it describes. It serves the social sciences as well as biomedical sciences and offers a useful range of examples of descriptive and inferential statistics.

3. *Introduction to Statistical Analysis*, 3rd ed., Wilfrid J. Dixon and Frank J. Massey, Jr., McGraw-Hill, 1969, 638 pp.

This is a good text for upper-level undergraduates and graduate students in the social sciences. It is clear and easy to follow, with numerous examples.

Index